Content With My Wages

A Sergeant's Story

Book III-Afghanistan

Also by the author:

Content With My Wages-A Sergeant's Story: Book I-Vietnam
Content With My Wages-A Sergeant's Story: Book II-The War On Drugs

Content With My Wages

A Sergeant's Story

Book III-Afghanistan

Gregory H. Murry
Master Sergeant, U.S. Army (Retired)

No End To Publishing Company
Austin, Texas, USA

Library of Congress Cataloging-in-Publication Data

Murry, Gregory H.

Content With My Wages-A Sergeant's Story: Book III-Afghanistan/Gregory H. Murry

Includes index.

ISBN- 978-0-9903976-3-2

First Printing

No End To Publishing Company LLC
Austin, Texas, USA 78715

noendtopublishing.com

Manufactured in the United States of America

And some serving in the military also questioned him, saying, and we, what shall we do? And he said to them, "Extort nothing from anyone by force, nor take anything by false accusation, and be content with your wages."
Luke 3:14

To

The members of ANA TAG II, the 1st Embedded Trainer Team, Coalition Joint Task Force Phoenix, and the young American men and women, who put down their smart phones and video games, raised their hand and volunteered to serve their country.

And To

The Afghan people and their National Army, the Afghan interpreters who risked their lives to assist us and when necessary, surreptitiously transformed our ill-considered declarations into culturally sensitive messages designed to bring a peaceful outcome to our encounters with the locals.

Khoda Hafez

Table of Contents

Combat Stress

The last thing the colonel did after kicking me out of the unit was to direct me to go to the Combat Stress Unit at Bagram Air Base. I was dismayed by the quick turn of events but this directive brought a brief involuntary smile to my face; what did he know about combat?

The next morning I joined the daily sick call convoy. We crossed Shomali Plain with its occasional knocked out Russian tank or armored car and small caravans of Kuchi with their camels and sheep. I knew we were close to Bagram Air Base when I could see the A-10 Warthogs on short final crossing over the road ahead of our convoy. After passing through the gate we drove down Disney Drive past the PX and pulled into the U.S. Army's medical facility. Across the parking lot was the U.S. Air Force Combat Stress Unit.

When I walked into the waiting room, a female Air Force sergeant signed me in and told me to take a seat and wait until my name was called. There were two guys from the 1-501 Airborne Infantry sitting there not saying anything. I picked up a copy of Air Force Magazine and absently thumbed through the articles while waiting for my name to be called. Finally the sergeant called my name and told me to go down the hall and enter the door on the left which I did. Sitting at a desk was an air force lieutenant colonel. He was thumbing through a medical record which I assumed was mine. His fingers were noticeably shaking and he seemed to be searching for some definitive reason for why I was here. After several minutes he looked up at me. He seemed startled but then told me to sit down.

I could sense a heightened level of tension in the room but I wasn't sure if it was emanating from him, me or both of us. He began to stutter until he managed to finally say, "Why are you here?" He wasn't looking me in the eye and now I realized that he was very nervous. "My CO sent me here," I answered. "Did he think you were a danger to others," he stuttered?

This is going to be fun I thought. "Do you think they would let me carry this pistol if I was?" I said, pointing to the M9 that was holstered on my hip. That seemed to push him over the edge as he began shoving his chair back while stuttering "Wha, wha wha?" I could see that he was suffering from his own combat stress, probably got it from his patients, and if I didn't do something quick this little trip dreamed up by the colonel was going to get ugly.

UZBEKISTAN
TAJIKISTAN
KYR.
CHINA
TURKMENISTAN
Amu Darya
Wākhān
(Wakhan Corridor)
Shīr Khān
Mazār-e
Sharīf
Kunduz
Nōshāk
HINDU KUSH
Toraghondī
Bagrām
Herāt
KABUL
Jalālābād
Shīndand
Ghaznī
INDIA
1972
Line of
Control
Indus
Kandahār
IRAN
Zaranj
PAKISTAN
0 100 200 km
0 100 200 mi
INDIA

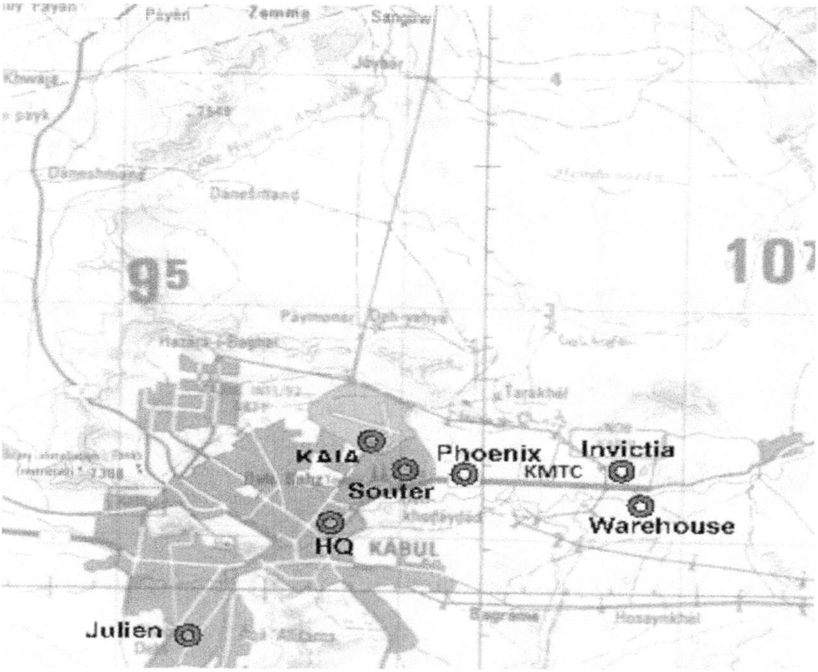
95
KAIA
Phoenix
Invictia
KMTC
Souter
Warehouse
HQ
KABUL
Julien

Author's Note

This is Book III of a war trilogy that I have written for a particular audience, the Infantry Sergeant and the Intelligence professional. If you're neither a sergeant nor infantry, read on at your peril. I began writing this narrative in 2006, after I returned from my third war, Afghanistan, and had retired from the Texas Army National Guard.

After the attacks of 911 and the subsequent invasion of Afghanistan resulting in the dispersal of the Taliban and their allies, a coalition of fifty-nine countries entered Afghanistan. Bulgaria, Canada, France, Great Britain, Mongolia, New Zealand, Romania and the U.S. were among the nations that joined together to form Coalition Joint Task Force Phoenix for the common purpose of rebuilding a national army for Afghanistan.

Joint Coalition Warfare is an intricate operational kabuki dance, with contention among the allies as intense as it is with their enemies. When the supposed masters of this chaotic environment, the U.S. Army Special Forces, were called away from this foreign internal defense task that is one of their core missions, the U.S. Department of Defense under Donald Rumsfeld decided to replace them with units of the Army National Guard.

For purposes of cohesion and unit readiness, the Defense Department took units from Texas, Rhode Island, and various other states and territories and put them under the command of a unit from Oklahoma and they added units of the Marine Corps, and Navy to the mix.

Our unit, TAG, composed of soldiers from Texas and Rhode Island, was responsible for assisting operations at the Kabul Military Training Center where the Afghan National Army was again rising from the ashes of the most recent conflagration that had consumed it. There I served as the Intelligence officer/NCO. Later I served as an assistant operations NCO with a team of combat advisors.

What follows is a cautionary tale of how two of those units, the Afghan National Army-Training Assistance Group II, and the 1st Embedded Trainer Team deployed to Afghanistan in the winter of 2003 and took up the challenge of rebuilding the Afghan Army. I have changed the names of some of the soldiers I served with and given them pseudonyms to spare them from any embarrassment that might ensue from my less than complimentary descriptions of their service.

Author's Note

What has been is what will be,
And what has been done is what will be done,
And there is nothing new under the sun.
Ecclesiastes 1:9

Preface

I wrote this to be a memoir, a history and a critique of the operations described. As a memoir, I am exposing my personal feelings about what I saw and did in Afghanistan. As military history, I hope to add to the body of information available for those who will study the effort made by the U.S. and its allies to rebuild Afghanistan's National Army; this wasn't the first time we have done something like this nor will it be the last. As a critique I'm not the right person to conduct this exercise; it was usually the job of the CO. I was, however, the oldest man in the unit and the only one to have fought in serious infantry contact, and had been an instructor and administrator of army leadership schools where by necessity and because of personal interest, I studied army leadership doctrine. I hope some others will take up the challenge to tell these stories, but until they do, here's mine.

In the past, combat operations were documented using battlefield interviews. These interviews were analyzed and became part of what is called doctrine. Training exercises were observed by umpires who made estimates regarding the effects of weapons and tactics and used these estimates to 'critique' the soldier's performance. In the aftermath of the Vietnam War, the U.S. Army was in shambles. The man most responsible for the tactics developed to fight the Viet Cong, General William DePuy, my Division Commander in Vietnam, would, ironically, ramrod the task of rebuilding America's Army.[1] Those charged with rebuilding the Army realized that the performance critique, used by leaders to correct their soldier's performance in training exercises, was no longer useful.

The introduction of the M.I.L.E.S. or Multiple Integrated Laser Engagement System made it possible for believable downrange feedback of the effects of rifle fire in training exercises without using live ammunition. Soldiers fired laser beams at each other and when they were hit, the device made a noise and they had to 'play dead.' To complement this technological development, the Army reached back in their history to the battlefield interviews conducted by S.L.A. Marshall and developed the AAR or after action review.

S.L.A. Marshall's battlefield interview technique was used by army historians in WWII, Korea, and Vietnam. The interview was done as soon as possible after the shooting stopped. The survivors were assembled in a group and questioned in a way that led the group to describe what actually happened in a particular action. This technique became a model for the modern AAR which was designed to eliminate the psychological dissonance that often resulted from the performance critique.[2] Now, according to army doctrine, the AAR would be, "a professional discussion of an event, focused on performance standards, that enables soldiers to discover for themselves what happened, why it happened, and how to sustain strengths and improve on weaknesses."[3]

All of this is well and good when things go ok, but the reality is that it is very difficult to get men and women to confront their weaknesses and failures. What started out as brutally honest AARs at the National Training Center (NTC) have degraded in many units into face-saving exercises usually conducted after a training exercise, if there's time, and very few battlefield operations are ever made the subject of a 'Professional Discussion.' General DePuy foresaw this and said in an oral history that "Performance oriented training is so demanding that weak sisters will begin to drift away from it… You can bet your farm on it!"[4] By the way, S.L.A. Marshall said that the starting point for understanding war is the understanding of human nature.[5]

In the military, it is natural to want to look good. From day one a soldier is taught to wear a clean uniform correctly and to always do what is expected. When he doesn't look good, a sergeant will immediately insure that the soldier takes corrective action. When the soldier performs poorly, once again a sergeant should be there demanding immediate corrective action. This causes soldiers to always strive to look and perform well. The difficulty comes in when the sergeants and the officers, who by virtue of their position, are also expected to always look good and perform their duties in a most excellent manner. When they don't their superior is presented with an ethical dilemma, usually of his own making. If leaders who know better are poor performers, the corrective measures available to the commander will usually adversely affect the career of the one being corrected.

This 'zero defects' mentality has been a part of the army since the 1960s. Before the all-volunteer Army, zero defects thinking could cost you a stripe or slow down a promotion; but if you showed your superiors that you had learned your lesson, you were back on track and moving along. Now with the all-volunteer force however, anything less than a perfect annual evaluation, much less a letter of reprimand, are considered career enders. This has created a climate in most units where mistakes by the senior people are covered up, which serves to demoralize the junior personnel while training them to do the same.

The leader should be the ethical and performance standard bearer of the unit and the unit members should be well acquainted with their leader's attitude toward professional performance. The troops' knowledge of their leaders' attitudes is developed by training together and getting feedback from AARs, counseling, and mentoring. This takes time.

The unit I deployed with to Afghanistan in the winter of 2003 had nothing in common with a unit preparing for combat. We were a military training institute with no wartime mission and most of the actual unit members didn't deploy. The commanding officer who was appointed to command us a month before we left, I didn't even know his name, was replaced by another officer shortly before we left. We were given a mission that had originated with the Special Forces and very little information on what exactly it was that we were supposed to do. We were pulled together from the largest and the smallest of the contiguous United States and we were commanded by a unit from a third state. Most of the people didn't know each other. The pre-deployment training only addressed combat subjects and political overviews of the situation in Afghanistan. As far a military doctrine goes, it was the perfect storm.

Since there were no AARs conducted during pre-deployment training nor were there any conducted in Afghanistan, nor any that I know of when the units returned, I am forced to conduct a performance critique like we did in the old days. For this I have asked a number of NCOs and officers to review the manuscript for accuracy and truthfulness; the mistakes and opinions, of course, are mine.

Acknowledgments

I have to thank the following people: Colonel Richard Orr (Retired) and Lieutenant Colonel Randy Davis (Retired) for reviewing this manuscript for accuracy. They reminded me of some events which I added or corrected.

CSM Brian O'Connors (Retired) for his corrections and suggestions; CSM Rogelio Hernandez Jr (Retired) and SFC Carl Carpenter for their feedback; Colonel Max Krupp SF for his review and documents.

There was another person who helped me through the entire ordeal of writing this story and desires to remain anonymous. For her I thank our God.

Acknowledgments

[illegible]

[illegible]

[illegible]

Abbreviations
(And other useful information)

AAR - After Action Report or Review
ACOG - Advanced Combat Optical Gunsight - Day/Night sight for rifle.
AGR - Active Guard and Reserve - Full-time Guardsman doing admin work
AK-47 - Russian Assault Rifle-7.62mm - An actual assault rifle.
ALOC - Army Logistical Operations Center - Forward Supply Depot
ALS - Automatic Link Establishment
ANA - Afghanistan's National Army
AN/PVS-7NVG - Single tube night vision goggles.
ARCOM - Army Commendation Medal
AT - Annual Training - Two week training period for National Guard
BAMC - Brooke Army Medical Center - San Antonio, TX
BDE - Brigade
BDOC - Base Defense Operations Center
BM-21 - Soviet MLRS
BMNT - Begin Morning Nautical Twilight
BMP - Armored Personnel Carrier, tracked, of the Soviet Union
BN - Battalion
BRDM - Armored Personnel Carrier of the Soviet Union
BTR - Armored Personnel Carrier of the Soviet Union
BUB - Battle Update Briefing - End of the day commander's briefing
CAB - Combat Action Badge
CD - Counterdrug
CDR - Commander
CG - Commanding General
CI - Counterintelligence
CIB - Combat Infantryman Badge
CJTF - Combined Joint Task Force
CJSOTF - Combined Joint Special Operations Task Force
CMB - Combat Medic Badge
CMIS - Counterdrug Management of Information System
CO - Commanding Officer
COE - U.S. Army Corps of Engineers
COL - Colonel - U.S. Army officer rank
CPX - Command Post Exercise
CSM - Command Sergeant Major - The senior NCO of a battalion or larger unit
CTF - Combined Task Force
D-30 - Soviet Artillery Piece-towed
DCU - Desert Camouflage Utilities
DD-214 - Official Record of Military Service
DEROS - Date Estimated Return Overseas

DFAC - Dining Facility
DOD - U.S. Department of Defense - The Pentagon.
DRA - Democratic Republic of Afghanistan-Communist Government
ERO - Engine Running On/Off Load
ETT - Embedded Trainer Team - Combat Advisors in the War on Terror
FTX - Field Training Exercise
FOB - Forward Operating Base
GO - General Orders
HESCO - Quickly established blast barriers that can be made into a fort.
HF - High Frequency Radio
HIG - Hezb-i-Islami Gullbuddin - Hekmatyar's organization
HQ - Headquarters - Military Commander's Location
HUMVEE - Wheeled vehicle that replaced the Jeep.
HVT - High Value Target - Leaders of the Taliban and Al Qaeda
IBAS - Interceptor Body Armor-Small Arms Protection - Bulletproof Vest
IED - Improvised Explosive Device - Command detonated roadside bomb
IG - Inspector General
Intel - Intelligence - A military term for useful information about an enemy
IPB - Intelligence Preparation of the Battlefield
ISAF - International Security Assistance Force - NATO peacekeeping force.
JDAM - Joint Direct Attack Munition-Smart Bomb
KBR - Kellogg, Brown, and Root - Company doing contract work for DOD
KHAD - The DRA Intelligence Agency - Like the Soviet Union's KGB
KMTC - Kabul Military Training Center
LRRP - Long Range Reconnaissance Patrol - Vietnam War era scouts
LRSD - Long Range Surveillance Detachment - Division level LRRPs
LZ - Landing Zone-Helicopter
M4 - Current U.S. Battle Rifle - latest variant of the M16 - 5.56mm
MBITR - Military Walkie-Talkie Radio - AN/PRC 148 multi-band
MFR - Memorandum For Record (CYA device)
MG-42 - German Machine Gun of WWII
MI-8/17 - Russian helicopter - The Hip
MI-24 - Russian Helicopter Gunship - The Hind
MOS - Military Occupational Specialty - An Army job description
MP - Military Police
MPRI - Military Professional Resources Inc - Contractor
MRE - Meals, Ready to Eat - U.S. military field rations
MTOE - Unit Organization and Equipment
NATO - North Atlantic Treaty Organization
NCO - Noncommissioned Officer-A sergeant
NCOES - Noncommissioned Officer Education System
NCOIC - Noncommissioned Officer in Charge
NDS - National Directorate of Security - Afghanistan's Security Agency
NGB - National Guard Bureau
NGO - Non-governmental Organizations

ODA - Operational Detachment Alpha - Special Forces A-Team
OER - Officer Evaluation Report
OGA - 'Other' Government Agency - If I told you…
OMC-A - Office of Military Cooperation-Afghanistan
OOGA - Other Other Government Agency - See OGA
OPSEC - Operations Security
OPTEMPO - Operations Tempo-The rate of military actions
PDSS - Pre-Deployment Site Survey (A Leaders Reconnaissance)
PEC - Pul-e-Charkhi - Suburb of Kabul
PKM - General Purpose Machine Gun - Russian made 7.62mm
POC - Point of Contact-Person to contact
PRT - Provincial Reconstruction Team
PSP - Perforated Steel Planking - Used for making forward airstrips
PT-76 - Amphibious tank of the Soviet Union
PX - Post Exchange-Retail shopping store on an Army Base.
QRF - Quick Reaction Force
REMF - Rear Echelon Military Forces or words to that effect
RIP - Relief in Place - A new unit takes over the duties of the old unit
RPG - Rocket Propelled Grenade - Russian anti-tank weapon
R+R - Rest and Recuperation
RTI - Regional Training Institute-Army National Guard training unit
S1 – U.S. Army staff officer responsible for personnel
S2 - U.S. Army staff officer responsible for intelligence
S3 - U.S. Army staff officer responsible for operations
S4 – U.S. Army staff officer responsible for logistics
SATCOM - A military radio utilizing a satellite repeater system
SATPHONE - A cell phone utilizing a satellite repeater system
SCUD - Short range ballistic missile of the Soviet Union
SF - Special Forces - The U.S. Army Green Berets
SITREP - Situation Report
SIPRNET - DOD internet used to share classified information
SOF - Special Operations Forces
SOP - Standing Operating Procedures
SPG-9 - Russian Recoilless Anti-Tank Gun
T-34 - The main battle tank of the Soviet Union during WWII
T-54/55 - The main battle tank of the Soviet Union in the 1950's
T-62 - The main battle tank of the Soviet Union in the 1960's
TAG - Training Assistance Group - American advisors to the ANA
TF – Task Force
TOC - Tactical Operations Center - Military Command Post
UCMJ - Uniform Code of Military Justice
VBIED - Vehicle Born Improvised Explosive Device
WFR - White Ford Ranger Pickup Truck - Small, 4 door, diesel engine.
ZPU-23-2 - Soviet Anti-Aircraft Gun - towed
ZSU-23-4 - Tracked Vehicle-23mm Anti-aircraft guns and radar.

Time

Times are designated using the 24 hour clock or military system of time. After 1200 (Noon) comes 1300 and so on.

Unit Abbreviations

1st Battalion, 179th Infantry - 1-179 Infantry
A Company, 1-179 Infantry - A-1-179 Infantry
1st Squadron, 4th Cavalry - 1-4 Cavalry
1st Battalion, 5th Artillery - 1-5 Artillery
Battery, Company and Troop letter designations are addressed as: A-Alpha, B-Bravo, C-Charlie, D-Delta

Ranks Abbreviations

Rank	Pay Grade	Abbreviation
Private	E-1, 2	PVT
Private First Class	E-3	PFC
Specialist	E-4	Spec
Corporal	E-4	CPL
Sergeant	E-5	SGT
Staff Sergeant	E-6	SSG
Sergeant First Class	E-7	SFC
Platoon Sergeant	E-7	PSG
Master Sergeant	E-8	MSG
First Sergeant	E-8	1SG
Sergeant Major	E-9	SGM
Command Sergeant Major	E-9	CSM
Sergeant Major of the Army	E-9	SMA
Warrant Officer	W-1	WO
Chief Warrant Officer 2	W-2	CWO 2
Chief Warrant Officer 3	W-3	CWO 3
Chief Warrant Officer 4	W-4	CWO 4
Chief Warrant Officer 5	W-5	CWO 5
2nd Lieutenant	O-1	2LT
1st Lieutenant	O-2	1LT
Captain	O-3	CPT
Major	O-4	MAJ
Lieutenant Colonel	O-5	LTC
Colonel	O-6	COL
Brigadier General	O-7	BG
Major General	O-8	MG
Lieutenant General	O-9	LTG
General	O-10	GEN

And I saw, and behold, another horse, a red one, went forth; and to him who sits on it, to him authority was given to take peace from the earth, and that men should slay one another; and to him a great sword was given.
Revelation 6:34

The Road to War

Afghanistan

Afghanistan[6] has a few things in common with Texas; it is about the same size with a similar climate and much of it is desert. In contrast, several mountain ranges much higher in elevation than those in Texas, run across it with many river valleys. Afghanistan is landlocked and its borders touch China, Pakistan, Iran, Turkmenistan, Uzbekistan, and Tajikistan.

The Afghans

The Afghans[7] and Texans also have a few things in common. They are both proud of their history and consider themselves a cut or two above the people who surround them.

The demographics of Afghanistan are difficult to describe statistically but it is generally assumed that the larger ethnic groups include the Pashtuns and Tajiks with the Hazaras, Uzbeks, and the Aimaq. There are smaller groups of Turkmens, Kazaks, Qizilbash, Wakhis, Nuristanis, Baluchis and Kyrgyz along with some Sikhs and Hindus.[8]

Nine or more languages are spoken with Pashto and Dari being the two official languages except in areas where the majority of the people speak one of six other languages.[9] The Pashtuns have dominated Afghan politics and all but one of their kings and most of their presidents have been Pashtun.

Afghanistan is the crossroads of central Asia. In the time of Ezra[10] in the Old Testament, when the children of Israel were released from their captivity in Persia and allowed to return to rebuild the city of Jerusalem and the Temple, only a remnant returned and the rest disappeared into history. Some claim that the Pashtun people of Afghanistan and Pakistan are their descendants which would account for their zeal in all things Pashtun.[11]

The first invasion of Europeans by the army of Sikandar, Alexander the Great, left blond hair and blue eyes in the gene pools of the people of this country. Many empires invaded Afghanistan; they rose and fell, but none seemed to want to stay. The Arabs came and left, but their new religion remained to trouble these people until now. Genghis Khan came and slaughtered many of the Afghans. He also left a legacy, the Hazaras, who still congregate in central Afghanistan in the Hazarajat.[12]

The Great Game

The modern history and the beginning of the nation we call Afghanistan began in the 1700s[13] shortly before the next wave of Europeans came toward Afghanistan, this time from the north, the west and the east.[14] The empires of Russia,[15] France[16] and Great Britain[17] were expanding and their struggle for dominance of the Indian subcontinent would come to be known as the 'Great Game.[18]'

Forward leaning British strategists saw Afghanistan as the buffer between each of these empires and the Russians saw Afghanistan as a route to India. Josiah Harlan, a man ahead of his time and the first American to come to Afghanistan arrived in 1838. A man who would be king, he would not be the last.[19]

After three wars with the British, the Afghans finally asserted themselves as masters of their own destiny. The British exacted their revenge by first establishing the Durand Line[20] as the boundary between India and Afghanistan. Then when they left India in 1947,[21] they created the nation of Pakistan with half its territory still claimed by Afghanistan. The European influence attracted some of the Afghans and an internal struggle between Pashtunwali,[22] the code of the dominating Pashtun majority, Islam, and democratic political theory gave new purpose to the Afghan pastime of tribal fighting.[23]

The Great Game II

By the middle of the 20th century, Afghanistan was once again in the middle of the 'Great Game.' The Soviet Union wanted the warm water port of Karachi which was in the new nation of Pakistan. With it they would control the entrance to the Persian Gulf and the world's oil supply.[24] A new player, the U.S. was now taking the lead in the struggle against the expanding Russian Empire by providing funds and engineers for agricultural projects in southern Afghanistan.[25]

Meanwhile the Russians seduced many of the Afghan intelligentsia and the Afghan communists began plotting the overthrow of the king and the establishment of the Democratic Republic of Afghanistan. The Russians sat across the Oxus, now called the Amu Darya River, and offered aid, building a road with a long tunnel through the Hindu Kush Mountains to Kabul.[26]

When the British drew the boundary between India and Afghanistan, they divided Pashtunistan,[27] the land of the Pashtuns, in half. Every Afghan king and politician had to make the regaining of the land from the border of Afghanistan to the Indus River in what is now Pakistan, their number one promise to the Afghan people if they wanted to be listened to.[28] The Russians promised to help them achieve these goals and the Afghans bought in. The Russians began training Afghan Army officers and pouring in military equipment.[29]

The king was overthrown in 1973 by his cousin, Daoud.[30] Daoud attempted to play off the various nations interested in Afghanistan, and when Saudi Arabia and Iran began competing for Afghan affection, he decided to greatly reduce Soviet influence.[31] The short lived Republic was eliminated along with President Daoud and his entire family in the communist led Saur Revolution of 1978.[32]

The new government of Afghan communists then tried to erase all previous Afghan culture, especially Islam, and the people in the countryside revolted.[33] The Soviet leaders watched this with dismay and in 1979, invited themselves in to help staunch up the newly established socialist state.[34]

On Christmas Day, 1979, the Soviet Union's 40th Army rolled across the Amu Darya on pontoon bridges, down the road and through the tunnel to Kabul. Other elements began air-landing at Bagram Air Base.[35] On December 27th, Spetsnaz, Soviet Special Forces, killed President Amin, because they suspected that he was beholden to the U.S.; the Soviets installed their own man[36] and attempted to subdue the Afghan nation before they moved on to Karachi and dominance of the Persian Gulf oil.[37]

For the next ten years, Afghanistan was embroiled in a vicious guerrilla war. Like the Russian and Chinese support of the North Vietnamese and the Viet Cong, the U.S. and a coalition of unlikely allies including Israel, Saudi Arabia and Pakistan, supported the Afghan resistance fighters.

The U.S. under the leadership of President Reagan, carried out one of the longest and largest covert operations in modern times. The Central Intelligence Agency provided arms, ammunition and financing to the Pakistan Intelligence Service (ISI) who channeled them to the seven mujahedeen, or 'holy warrior' groups that were doing the fighting in Afghanistan so that they could bleed the Russians with the 'death of a thousand cuts'.[38]

On February 15, 1989 the last soldier of the Limited Contingent of Soviet Forces crossed the Amu Darya River and the war in Afghanistan entered its next phase, the civil war of the Mujahedeen versus the Democratic Government of Afghanistan. This lasted until 1992 when the Najib government fell and a coalition government of mujahideen ruled the country for a short period of time before resuming the civil war, this time between the various mujahideen factions.[39]

The Taliban

In 1994, a new force appeared in southern Afghanistan, the Taliban. Supposedly constituted by religious students, Pashtun mujahideen fighters encouraged and supported by Pakistan began taking control of the country, province by province, establishing a Sharia government and sweeping aside all opposition as they moved north.[40]

Most of the Pashtun mujahideen joined them and the rest of the Afghans formed the United Front; ethnic groups of Tajiks, Uzbeks, and Hazaras fighting under the leadership of Ahmad Shah Massoud. They made a stand in Kabul until 1996. When Kabul fell to the Taliban, the United Front retreated to the Panjshir Valley. The Taliban continued north and by 2001 they controlled most of Afghanistan.[41]

The Arabs-Al Qaeda

There were other forces at work within this coalition. Ever since the Six Day War in June of 1967, when Israel defeated Egypt, Jordan, and Syria, the Muslim world had been in turmoil.[42] Out of this turmoil came men with a vision of a new pan-Arabic nation that would eventually control all the nations of the earth, the restored Caliphate.

These men were reformists who followed the teachings of Ibn Abd Al-Wahhab. These teachings had influenced groups such as Islamic Jihad and the Muslim Brotherhood of Egypt, and were the foundation of the Saudi Arabian version of Islam.

Their basic message was that the ummah, or community of Muslims had been contaminated with western culture and that in order to gain the blessings of Allah, they had to cleanse themselves of this culture. Having realized that this cleansing would not take place spontaneously, they declared themselves the vanguard of this movement and set out to overthrow the governments of traditional Islamic countries like Egypt in order to establish Sharia or Islamic religious law and government. They assassinated Anwar Sadat, the president of Egypt, after he made a peace treaty with Israel.[43]

When the Soviet Union invaded Afghanistan this vanguard rallied support for the defense of Islamic territory, a Jihad, and joined forces with the Afghan mujahideen. They fought alongside the Afghans and formed relationships. After the Russians left they returned to their own countries where they found themselves unwelcomed and eventually many of them returned to Afghanistan where they set up training camps to prepare the next generation of Mujahideen for their vision of many Afghanistans.

Among these Arab's were Osama bin Laden and Ayman al Zawahiri who began establishing a base, an Al Qaeda, in Afghanistan for these efforts. When the Taliban rose up they joined them. While the Taliban had a limited vision of establishing Sharia in Afghanistan, Al Qaeda had a world view and plotted to attack what they perceived as the source of the contamination of Islam, the United States.[44]

911

On September 9th, 2001, the leader of the United Front, Ahmad Shah Massoud, was assassinated by the suicide bombing myrmidons of Osama bin Laden[45] and two days later some more of them hijacked four passenger airplanes in the U.S. and flew two of them into the twin towers of the World Trade Center in Manhattan, New York City. Another one was flown into the Pentagon and the fourth crashed in Pennsylvania after the passengers fought back, the first counterattack of the war on terror. For the record, none of the hijackers were Afghans. The U.S. Intelligence community determined that Al Qaeda was behind the attack and President Bush demanded that the Taliban turn Osama bin Laden over to the U.S.[46]

Pashtunwali

The Pashtun Taliban, apparently bound by Pashtunwali, the way of the Pashtun, which required them to give refuge to anyone who asked for it, would never agree to these demands[47] so the U.S. entered Afghanistan at the 'invitation' of the leaders of the United Front. First a CIA team, led by a 60 year old man who interrupted his retirement to be the first in, and then together with teams of U.S. Army Special Forces, they conducted a classic special operation combining cash, precision air-strikes and tribal warriors on horseback.[48]

Nation Building

By December of 2001, the survivors of the Taliban and Al Qaeda had fled to Pakistan and the U.S. with a coalition of NATO nations and others, set about to establish a stable government in Afghanistan. There had been several large actions in places like the Tora Bora area of the White Mountains on the border with Pakistan[49] and in the Shahi Kot Valley (Operation Anaconda),[50] also near the border with Pakistan, where U.S. and Coalition forces attempted to destroy any remaining Al Qaeda and Taliban remnants in Afghanistan. After these headline making actions, the world didn't hear much about Afghanistan as the media focus shifted to speculation on another war with Iraq. An interim government under Hamid Karzai was established and progress was steady until the U.S. invaded Iraq in the spring of 2003.

Nonetheless, the coalition forces that remained in Afghanistan were busy. The first priority was security. The Afghan National Army was being reconstituted under the leadership of the United Front, now called the Northern Alliance. The Kabul Military Training Center was established on the east side of Kabul where the old Afghan Military Academy had once turned out professional officers; now the Afghan officers, NCO's, and soldiers were all being trained here by the French, British, and Americans.[51]

My Road to War

My own road to this war began the day President George W. Bush told the Armed Forces to "get ready!" On September 11th I had watched the TV news people replay the money shots over and over; I told another sergeant that this was his Pearl Harbor; but when the President spoke I realized it was mine also.

Like many other members of the armed forces I spent most of my free time staring at the pretty girls working for CNN and other news outlets as they introduced us to a new generation of war correspondents who used satellite technology to report in real-time from northern Afghanistan. That fall I wore out my TV remote flipping through the news channels trying to figure out how or if the National Guard would have any part of this war.

The Vietnam War was famous for leaving most of the National Guard combat forces at home.[52] Then came the Gulf War; all the National Guard 'Enhanced' Brigades that were meant to 'round-out' the regular Army divisions were left behind to conduct some of the longest NTC rotations in history, while regular forces were cannibalized to provide the manpower to bring the deploying regular Army units up to strength. There had been a lot of brouhaha about that after the war was over but I still couldn't see a role for us in this one.[53]

Then there was a glimmer of hope on the horizon. An old comrade from the Long Range Surveillance Detachment called me to ask if I would be interested in going to Afghanistan with his unit, the Information Operations Group. It didn't sound like a fix bayonets and charge type organization but, then again, with that kind of name, who the hell knew what they did?

I said yes immediately and went home with a happy heart. When I told my wife that it looked like the Guard was going to send me to Afghanistan, she said "If you don't get killed there, I'm going to kill you when you get back." I decided I should fellowship with one of the elders of the church. When I asked brother Don how he felt about my going to Afghanistan, he stood there for a long moment. I was starting to fidget when he said, "Well Greg, I would feel very lowly about that." I was crushed but I took that speaking as from the Lord. The next day I called my friend and after apologizing to him asked him to take my name off the list. Then I sat there and wondered what the Lord had in store for me in the Great War on Terror.

The Counterdrug Task Force

In the summer of 2003 I was the On-Call Operations Sergeant for the Counterdrug Task Force (CD), tracking the missions of the special ops and aviation elements. I was also deeply involved in the reorganization of the Task Force and the re-classification of 100 cargo inspectors who were now going to be intelligence analysts for the DEA.

I had been on CD active duty since June of 1989. For the first ten years I was the operations supervisor for the Special Operations Detachment (SOD), an ad hoc special operations unit doing surveillance in direct support of any law enforcement agency conducting narcotics interdiction and investigations. We were always armed and usually in civilian clothes.

We traveled continually to diverse places from the Red River to the Rio Grande and from El Paso to Beaumont. Most of the operators came from SF and Ranger units but we put them all through our own operator training course before they could work with us.

Our standards were high; the training was intense; our planning was meticulous and mission execution was usually flawless. This had conditioned me to view service in the National Guard very differently than most of the other soldiers in the Task Force and in my other job as a traditional Guardsman, a 'weekend warrior.' Although we were doing great things, none of our efforts counted for promotion purposes. We had to be a member of a traditional National Guard unit and drill with them one weekend a month and attend annual training with them for two weeks every summer.

The 136th Regional Training Institute

On drill weekends I was assigned to the 136th Regional Training Institute (RTI), the Academy, at Camp Mabry in Austin, TX. My membership in the Texas National Guard was based on my assignment to this unit. The Academy ran the Officer Candidate School, two levels of the Non-Commissioned Officer Education System (NCOES), and numerous combat arms training courses. I was the Branch Chief for the Advanced Non-Commissioned Officer Course (ANCOC). We had just graduated a class and were finishing up the paperwork for the student records. The class had gone well; my superiors were happy and I was bored.

For also if the trumpet gives an uncertain sound,
who will prepare himself for battle?
1 Corinthians 14:8

Chapter 1: August-November 2003

A Rumor Of War

In the late spring of 2003 on the east coast in Rhode Island, the 243rd RTI was preparing to relieve the Vermont RTI in Afghanistan. The Vermont RTI had replaced the 3rd Special Forces Group and was the first National Guard unit to deploy for the mission to rebuild the Afghan National Army.

Vermont's entire RTI had deployed and problems had arisen, so after some reconsideration, it was determined that sending the entire Rhode Island RTI to Afghanistan would adversely affect the accomplishment of that state's training goals. A call went out for another RTI to combine with the 243rd and that's how Texas became involved.

Since the Texas RTI was commanded by a full colonel, command of the expedition was given to Texas.[54] Of course we knew nothing of any of this as we merrily went about our annual training routines and prepared for the next NCOES classes that would begin in the fall.

During that summer of 2003 I began to hear rumors of a mission to Afghanistan for the Academy. I went to Terry, the unit admin sergeant and asked him what he knew. He told me that the Regiment might be going to Afghanistan to train the Afghan Army. He said a senior officer of the unit had heard that there was a call for a National Guard Academy to run an Afghan training facility and he had suggested that Texas was ready. Since training foreign armies is a Special Forces function, I thought that this was a bunch of baloney.

The rumors, however, persisted. Terry told me they were starting a list of volunteers and, did I want to be on the list? I was 57 years old and out of shape. I had just bought a new house and my son Richard was a junior in high school. The CD program would not allow us to volunteer for other duties; so I told him that I couldn't volunteer but that I would be very disappointed if they didn't order me to go.

Several days after this I ran into a senior officer of the unit in the cafeteria. I'll call him Colonel Knight. I told him that I had heard that there was a possibility that the Regiment might be going to Afghanistan. Colonel Knight acted coy but said that it was possible. I informed him that because I was with CD I couldn't volunteer, but that I would be very disappointed if they didn't take me. "I was born for this mission," I told him and he said something to the effect that he wished that the other senior NCOs in the unit felt the same way.

I found out later that he was the one who had heard about the mission and had volunteered us. I started to bother Terry every few days to find out if he had heard any more. He told me that I was being considered for the senior NCO job with the deploying group.[55]

I was excited. Be careful what you wish for, right? Good advice, but it had never worked for me; having one more chance to go to war before they ran me off overruled any sense of foreboding that I might have had.

Their considering me for a position meant that they might order me to go instead of playing the volunteer game that seemed to be so popular in the National Guard. I guess the Guard leadership thought that if they had volunteers to fill up units that were being mobilized, then they could tell the American people that the Guard could perform their missions with an all-volunteer force. Very few members of the 136th Regiment had volunteered to go to Afghanistan and I wondered how they were going to get the people they needed if the unit was mobilized to go.

The SRP

At the end of August the RTI was notified that they would go through the Soldier Readiness Processing (SRP) on the 5th of September to prepare for mobilization.[56] It looked like the rumors were true after all. We went out to Camp Swift, a National Guard training area about 30 miles east of Austin in Bastrop. The SRP was conducted there and it took all weekend.

I was sitting on a bench waiting to see someone at one of the stations when I heard the guy sitting next to me say in a plaintive voice," We're the Academy; we don't go anywhere." That was true up to this point.

The Academy's wartime mission was to go to Fort Hood and train soldiers in the schools run by the regular army so that the regulars could be freed up to fight. I told him that it looked like our mission had changed and smiled. He looked at me like I was crazy.

The Uncertain Trumpet Sounds Again

The next day, Sunday afternoon, we gathered in RTI's auditorium and the regimental commander took the stage to tell us about the mission. He started out by telling us that it was voluntary; that we should all consider how going might affect us; that we should not go if it might have negative effect on our lives.

After these words of inspiration he then told us that he had a heart condition that could prevent him from going. He also told us that he could have a surgical procedure that would correct the problem but that the doctor told him that he could elect to have the procedure later in his life if he chose. He said that he would have the procedure later and that he wouldn't be going.

This was a really inspirational talk from the commanding officer of a unit that was being mobilized for the Great War on Terror. Here he had advanced to the rank of full colonel in the National Guard, but when a war came he wouldn't be able to attend.

As the colonel went on with his soul-stirring words, I thought about our regimental history in past wars. The Academy had recently changed its lineage under the Army system of unit designations. We had worn the unit crest of an artillery regiment whose motto inscribed on the crest was "We Play the Game."

I hated that motto because it was a popular term in the pre-Vietnam army and I blamed that kind of thinking for our loss of that war. The unit's glorious history during WWII included four years of POW slave labor in the jungles of Burma working on the railroad made famous by the great movie, "The Bridge over the River Kwai." I used to wonder if they played "The Eyes of Texas Are Upon You" at their reunions; it having the same melody as "I've Been Working on the Railroad.[57]"

Our new unit designation was the 136th Regiment and our crest proclaimed "Tidal Triumph" a reference to the two amphibious assaults conducted by the 36th Infantry Division in WWII. Salerno was the first and the history of that operation has been spun every way possible. It was the 36th Division's baptism by fire and by all accounts it was a rough one. The second operation was in Southern France and went much better, the division having picked up a lot of experience in Italy. I liked the new motto better than the old one because I could connect it to surfing.

As I sat in the auditorium listening to his 'rally round the flag' speech I wondered why I had told them that I wanted to be ordered to join this expedition. At the end of the SRP I knew that there was a way out of all my bravado because I too had a problem. I was told that I was in a non-deployable status because I had a tooth that needed a root canal.

The colonel's speech had its effect and there were a lot more people from the Regiment who would not be going. The Command Sergeant Major was unable to deploy. The Operations Sergeant Major would not be going; the operations sergeant who had been awfully quiet during the SRP would not be going. The only full time personnel who wanted to go were the senior officer who had got us into the mission in the first place and Terry the admin NCO. After the meeting Terry told me that I would be ordered to go.

That Sunday afternoon I handed the ANCOC operation over to George and Dave. George was senior so he became the Branch Chief and Dave remained the Senior Small Group Leader. I gave them all the handbooks and SOPs I had developed over the past four years and said goodbye to NCOES. I was going to war and as much as I hated going to the dentist, I still called and made an appointment on Monday morning.

The New CO

The RTI got a new commander shortly after the SRP. The new CO was a full colonel; he wore armor branch insignia and had been LTC Randy Davis's armor battalion commander. LTC Davis and I had served together in the CD Task Force for many years. He said that the new CO was a West Pointer and had been on active duty for a long time. Now he was a part timer and worked for a defense contractor. He also said the colonel was a good officer and a man of integrity. One day at drill LTC Davis introduced me to the new CO. We talked for a short time. He didn't have a lot to say to me so I let him know that I was here to help him accomplish the mission and told him to let me know if he needed anything.

PDSS Comes Before PTSD

Colonel Knight and Major P were alerted to go on a pre-deployment site survey (PDSS). They went to Afghanistan for about a week and came back with a lot of information about the mission.

Upon their return Colonel Knight briefed the command group.[58] Our mission would be to provide training assistance to the Afghan National Army (ANA) at the Kabul Military Training Center (KMTC). The 3rd Special Forces Group had originally had the mission, but the Special Operations Command felt that they needed to get their personnel back to conducting more direct action type missions. They requested that their troops be relieved of these responsibilities and be replaced by conventional troops.

At first the 10th Mountain Division had the mission but then it was determined that what better type of unit could conduct this type of mission than a National Guard Regional Training Institute. The Vermont National Guard RTI was given 30 days' notice to prepare themselves for this mission and then they were deployed. Like the guy said in our SRP, "We're the academy; we don't go anywhere!" When they arrived in Afghanistan, all they had for weapons were 9mm pistols.

The Plot Thickens

More deployment information was starting to come in. We would be joined by twenty-four soldiers from the Rhode Island National Guard and formed into the Training Assistance Group (TAG). We would be part of the coalition's Combined Joint Task Force Phoenix which was led by the 45th Infantry Brigade of the Oklahoma National Guard. Our role would be to operate KMTC and provide training assistance to the ANA basic training component. We would be working with French and British who were responsible for training the ANA officers and NCOs.

The 'Battle' Of The Bone Yard

Colonel Knight told us that there was still combat activity near KMTC and that a battle had taken place there recently that was known as "The Battle of the Bone Yard."[59] Some of the Americans including LTC K from Rhode Island, who was on his own PDSS, were returning to KMTC from the range when they were fired on from some abandoned buildings. They chose to stop and when they received more fire, they returned it and the battle began.

One of the LTCs was hit twice in the chest but his body armor prevented him from getting a possible fatal wound. The group exchanged fire with the people in the buildings for a time and eventually reinforcements arrived and they were able to capture seventeen of the people that were shooting at them. There weren't many more details but Colonel Knight said that we would be advising the KMTC security force which hadn't been done up to that time. Then he said we would be sending our people out with them each night to conduct combat patrols around the base. This alarmed me. The RTI was a training academy not a combat unit. I told him as much and recommended that they find some combat soldiers to perform these duties.

He said that there were none and we would have to do it. In that case, I told him we would need to do a lot of training as a unit and that we should get some special equipment like thermal image scopes, radios, long range sniper rifles, etc. I wrote up a risk assessment worksheet to put my concerns on paper using the current army method for describing the components of dangerous activities and what could be done to lower the risk.[60] I gave it to him and he said he was impressed with my analysis of the task and told me I could put together an equipment list and he would submit it.

Just Because You're Paranoid...

Something wasn't right but I couldn't put my finger on what it was. I had been under the impression that combat operations were over in Afghanistan. There were no news reports about firefights in the Kabul area. My paranoia began to kick in as I listened to a West Point LTC telling us the real situation. I looked around at the people who were going and I began to get a sick feeling in my stomach. I was going into infantry combat with the Texas National Guard Academy. I really wanted to go on this deployment but I didn't want to get killed in some National Guard fiasco just before I retired.

I emailed Captain Brian S my former CO at SOD. He had gone to Fort Hood as soon as he heard of a possible Iraq deployment for the 1st Cavalry Division. That didn't work out but he managed to get a job as a combat advisor and was in Afghanistan. I asked him if he had any info about what was going on around Kabul but he was far from the city and when he replied he said that I would get all the information I needed from the PDSS team.

I went to work and spent hours on the internet trying to find the latest and greatest equipment and the national stock numbers for them. I continued to voice my concerns to Colonel Knight regarding unit training to prepare for nighttime combat patrols with foreign troops who didn't speak English. This became a source of tension between us because I felt that he was being too casual about the risks involved with his plans. He did not have much experience conducting night operations using the current technology. He told me that we may have to do these patrols with no modern equipment because the ANA didn't have any.

I argued that it wasn't worth our taking any casualties that could be prevented by the use of technology that was currently in the army system. The more we argued about these things, the more worried I became. Were we heading into a possible disaster because our leaders were looking for some excitement rather than just accomplishing the mission? We weren't getting any intel about the current situation at KMTC so I assumed that incidents like the "Battle of the Boneyard"[61] were taking place every night, although there was nothing in the news, which was now dominated by the war in Iraq.

Volunteers of America

The lack of available RTI personnel caused the Academy leadership to seek volunteers from the other Guard units in the state and even out of state. People from different units came from all over the state. I didn't recognize any of the names on Terry's growing list. There was a strong element, at least 3 officers and a number of sergeants, from one of the field artillery battalions.

I was looking down the list one day with Terry and I saw SFC Roger Hernandez's name on it. I was delighted. I had known Roger for several years in SOD and had put him through OTC as a member of the largest OTC class we had ever had. Roger had been an excellent operator and was now a team leader. He was transferring to San Antonio to become an intelligence analyst with DEA. I called him and told him that I too was going to Afghanistan and would he like to come to Austin a few days before we were mobilized and help me order equipment for the mission. He said yes and I had him put on orders.

I did a quick area study of Afghanistan on the internet and noticed that it got very cold there in the winter. This began a quest for cold weather clothing and equipment which didn't really end until we left Fort Carson. I hadn't served in a cold weather since I left Germany in 1965 but I sure didn't want to freeze to death over there because of a lack of gear. I bought a large rucksack, an extreme cold weather sleeping bag, lots of long underwear, socks, gloves, hats etc. I was doing anxiety buying and it wasn't helping. I also bought several video cameras and a new laptop computer. I figured that we were going to make some extra money over there and that I would be covered.

We were mobilized in November. Roger and I had been working on the equipment list for a week and now the other people began to show up. Some of Roger's buddies from his artillery unit were down in Peru training with the Peruvian army. They flew back to the states, went home for a night and showed up at Camp Mabry the next day, ready to go to Afghanistan. These guys were hooah!

We had a staff meeting and Colonel Knight briefed the CO on the current information regarding the mission. When he got to the part about the security mentoring mission and how we were all going to be conducting combat patrols each night with the ANA, the CO looked at him like he was crazy. Later in the office I told the CO that I thought the patrols were a bad idea.

It's Not Your Fault If You're Outranked...

I was pretty frustrated with Colonel Knight by this time and told the CO that Colonel Knight went on the leaders recon and when he came back he couldn't show me on a map where it was we were going. I told him that Knight appeared to be very unprofessional and that the CO should consider some serious unit training if we were going to do combat patrols with a foreign army that didn't speak English. He told me that he wanted me to say the same thing to Colonel Knight in his presence. I was surprised at this because as a sergeant I had no way to prevail in an argument with an LTC.

The most frustrating aspect of this was that I was making assumptions based on information provided by Colonel Knight who had lost his credibility with me some time ago. Now I was being made to look like I was afraid to go into combat. I read somewhere that we all have to die but we didn't have to die stupid. Colonel Knight's cavalier approach to what he described as nighttime combat patrols with a mix of Afghan soldiers and Texas National Guardsmen of every MOS who had not trained in this specific skill set sounded to me like stupid.

The CO called me into a meeting with Colonel Knight and told me to state why I thought his proposals were a bad idea. By then I had determined that Colonel Knight was a bit of an idiot. I knew that I couldn't prove it so rather than expose myself as being in the same status I tried to explain my objections. Having untrained members of our thrown together circus going on nighttime combat patrols with people who didn't speak English and without the equipment currently being used by combat units in that country, seemed extremely high risk. The only way to mitigate these risks was an extensive training program and special equipment. He asked me to elaborate on the equipment so I told him about thermal image systems that would allow us to detect intruders five miles away in total darkness and smaller systems that could be hand carried. I also mentioned over-watch tactics using long range sniper rifles with thermal scopes.

It Is Your Fault If You're Outsmarted

I told him of my concerns regarding the fact that we were not an infantry unit. If we were going to conduct nighttime combat patrols where, according to Colonel Knight there would be a high probability of contact with Al Qaeda or the Taliban, then we needed to set up a training schedule that emphasized light infantry tactics and marksmanship along with the use of night vision equipment. Finally, I said that if we didn't do these things then we should request an infantry unit be assigned to us as our security force. "We shouldn't be doing these kinds of operations without the training and equipment," was my final word.

The CO then said he disagreed with me and that we would do what was necessary to secure our area of operation. If he had his own intel, he wasn't saying. Colonel Knight was sitting there with a smirk on his face so I left the office feeling like I had been outsmarted by this guy and now I would be put in harm's way by idiots. I did write the CO a long memo with my recommendations for pre-deployment education and training but events were moving faster now and we could only react to someone else's time schedule.[62]

The Search For An Advantage

I did have permission to continue to find equipment that would be useful. I talked with LTC Randy Davis about bringing one of the LORIS systems that we had used with great success in the detection of drug traffickers five miles away. He looked into it but we were running out of time and no one could figure how we could get permission to take one out of CD and into the War on Terror or whatever they were calling the invasion of Afghanistan.

Roger and I searched the internet and found information on the Air Force Thermal Image System that they used for air base defense and the National Stock Number (NSN) for it. We found some small thermal devices like we used in SOD and we found the latest and greatest radio system, the MBITR then being used by SOCOM units. It would talk to a variety of older radios and it had good range so we put 40 units on the wish list along with several .50 caliber sniper rifles and some thermal scopes and thermal ID devices that could be worn by the people on patrol.[63]

We gave the list to Major P who said he would do what he could to get the things we asked for. He was an infantry officer with a Ranger tab and I think he was empathetic with us but as a major he had to watch his step around Colonel Knight and the CO. I got Roger to take a long road march with me around the camp. I had hoped to get the CO interested in doing some pre-deployment training but I guess he was starting to consider me a crazy NCO and wasn't very open to my suggestions. I can't blame him because I was in my combat zone mindset which I had been in for the past 10 years in SOD and I forgot one of Bil's early lessons about Spartans in an Athenian camp.

My Funeral Plans

I did what I could to get my family ready for my deployment. In typical infantry manner I was joking when I told my son Richard that if I got killed I wanted him to make the funeral arrangements and have my body face down in the coffin with my pants down so that whoever came to my funeral could kiss my ass. I thought it was kind of funny but he was young and took it more seriously and went into a time of depression which became obvious when he started failing all his classes after I left.

A New House

We had brought the house next door and were having some remodeling done by one of the brothers in the church. He was making suggestions on how we could make the home more useful for home gatherings and soon we were in some major surgery on the house but the end result was quite good. My wife, however, was having a hard time with everything and when it came time to move in she resisted. There was an addition to the back of the house that the previous owners had built for a workshop so I moved all my video equipment and computers back there. She was still resisting as the time became closer for my deployment so I invited the young people from the church to have their Saturday night meeting in the big room; it accommodated 35 people so I was happy.

Time For The Circus To Leave Town

The time finally came for us to assemble at the academy for three days before we got on a plane and flew to Fort Carson, Colorado for pre-deployment training and unit validation before we flew to Afghanistan. I had arranged for several of the Information Operations people who had recently returned from Afghanistan to give us a briefing on what they knew about the current situation and they gave a fairly good briefing about the history and future trends. They did not mention any recent combat actions in the Kabul area so I asked each one of them privately to give me the no-shit truth about what was going on there and they said that nothing was going on, it was quiet.

Of course, this didn't square with Colonel Knight's report about what was going on there so I was once again thrown into turmoil about what we were getting into. If it were quiet then Colonel Knight, now a senior officer of this lash-up was going to be a big problem over there. He was obviously crazy if he would invent situations that didn't exist. If it weren't quiet, like the IO guys claimed, then we were being lured into a bad situation that the government and the press were conspiring to conceal from the American people.

The Sergeant Major

I was still the senior NCO and Terry told me that I was to be the sergeant major for the whole unit.[64] I wasn't much interested in that because I didn't really know what a sergeant major did. I took care of the few formations we had, got the CO to get rid of the berets and order us to wear the soft patrol caps, and made the point several times for everybody to take care of any family problems before we left.

When the day arrived for us to leave I showed up with my duffel bags and rucksack and got on the bus with everyone else for a trip to the airport. I had a small video camera and I used it to take video of almost everything we did from here on until we came home.

Fort Carson

We arrived at Peterson AFB in Colorado Springs and were bussed to Fort Carson. At an elevation of 5800 feet it was almost the same elevation as Kabul, Afghanistan. This made for perfect acclimation for our mission. Many of the units going to Afghanistan and later to Iraq, deployed from places like Fort Polk, Louisiana, a pine tree swamp, or Camp Shelby, MS, another low altitude briar patch.

Fort Carson had already prepared the 45th Infantry Brigade from Oklahoma, our parent unit for this deployment and now, almost as an afterthought, they were preparing us to go. The first thing we did when we got there was to meet the Rhode Island contingent. These soldiers were like us, ostensibly from the Rhode Island RTI, they were actually from all the units in that state and a few were from surrounding states, Rhode Island being the smallest state in the USA.
The first thing I noticed about them was the way they talked. They said 'youse guys,' while we said 'y'all.' A captain from Alabama also joined us; he just appeared at Fort Carson when we were doing our pre-mobilization training.

The Army Reserve has units that conduct pre-deployment training. There was an NCO from an Army Reserve unit assigned as our liaison who was a total loss. He didn't know where most of the places were that we had to go to and he wasn't very interested in finding out.

I managed to get a map of Fort Carson with building numbers and this helped us get around the first few days.

We tried to organize the unit into the sections we would work in when we got to Afghanistan but since we really didn't know what we were going to do, we appointed squad leaders by rank for accountability purposes and muddled on. Since the lowest ranked individual was an E-4 (and she was the only one) and the rest were mainly senior NCOs and field grade officers getting things done was like herding cats until the natural leaders made themselves known and started assisting the ones in charge.

The CIF

The pre-deployment checklist was long and detailed. Every day we would move from place to place at Fort Carson checking the blocks that brought us closer to an airplane that would fly us to the war. One of the first places we went was the CIF or Central Issue Facility. This was an all-day event where we filled large shopping carts with everything from hats to boots: cold weather sleeping bags, bear suits, and web gear. It took me two duffel bags to carry everything out of there. One of the items was the ballistic vest or Interceptor Body Armor-Small Arms Protection (IBAS). It weighed about 30 pounds and was very cumbersome. I soon found out that this was not an optional fashion statement; we would have to wear them all the time. We were issued several pairs of cold weather boots but the best ones for me were the Bellevilles with Thinsulate. When we checked out we had to sign a huge green computer printout because when we returned we would have to give them back all of these nice things. I was happy to sign because I felt in signing it that the Army was going to bring me back so I could turn this stuff in.

Task Force Martin

Colonel Knight held a meeting with the senior NCOs and proposed we call ourselves Task Force Martin[65] after a Rhode Islander named Albert Martin who died at the Alamo. No one had any objections so that's what we were called by someone, somewhere. Eventually a unit coin was made with the same name but I don't think anyone gave it a great deal of thought except for Colonel Knight. We had a soldier from Texas who was a direct descendant of one of the men killed at the Alamo. Colonel Knight had found him somewhere and convinced him to volunteer. I soon determined that this young man was suffering from some kind of mental strain and after a few incidents of strange behavior on his part the CO saw the same thing, agreed with my assessment and sent him back to Texas.

The S2

After a few days I was told to report to a meeting with Colonel Knight. He told me that all the senior members of the command group were from Texas and that we needed to include someone from Rhode Island. Since the two officers were not giving up their positions, that left me. Colonel Knight asked me what I thought and I had nothing to say. I wasn't that interested in being the sergeant major anyway. There was a master sergeant from Rhode Island at the meeting and I figured that he was the one they wanted. Colonel Knight told me I could discuss the situation with the CO but I chose to do nothing and to see what would happen next.

The next day I was told that I was now the S2, the intel guy.[66] Since I was the only intel qualified person in the unit, I would occupy an officer position on the staff. I had no idea what the S2 would do on this operation but I began my duties by scanning the news every day and getting a weather report from Kabul. There was a staff meeting each day and I would give a short briefing on these subjects.

I looked around Fort Carson for some more intel resources and found a sergeant who had recently come back from Afghanistan. He agreed to give a briefing to everyone with a secret clearance. When we assembled in the secure facility he gave us a rundown on the "Panjshir Valley Boys" who were currently running the country and were better known as the "Northern Alliance."

This was interesting background information but at the time we were being inundated with so much 'information' that these details were lost on most of us. Since we would be working with the ANA, the relationships to the personalities discussed here and their connection with the officers we would work with on a daily basis were important but would be difficult to discern.

We also received briefings from the U.S. State Department regarding the importance of our mission which was to assist the ANA leadership in the conduct of training at KMTC. The building of the ANA was described as our exit strategy from Afghanistan so we were all somewhat inspired by the importance of our mission.

We got a briefing on Afghan culture from a Pakistani Brigadier General. How he knew so much about Afghanistan and why he was here in the U.S. telling us about these things, caused me to believe that he was from the Pakistani intelligence service (ISI) which had been our partner in the Jihad against Russia and which was now playing a double game with us.

I kept waiting for the real deal about what was actually going on at KMTC, still being concerned that we would all be engaging the enemy in firefights around our base. Everyone that I asked about the current level of violence said the same thing; there was nothing going on in Kabul.

I finally got a little clearer about what was going on in Colonel Knight's mind. One of the sergeants who had been assigned to advise the Afghan security element at KMTC told me that Colonel Knight had told them that they all might be able to get the Combat Infantryman Badge (CIB) if everything went well. Now I got it; he was on a badge hunt. As it turned out he wasn't the only one in Afghanistan that was on the hunt for badges and colored ribbon that year.

The CO's Command Philosophy

During one of the many briefing sessions, the CO passed out a two page document entitled "ANA-TAG Command Philosophy."[67] It described our mission, the end state or results of our deployment and the operational concepts he wanted for us to keep before us as we carried out our duties, duties that were still pretty vague in their details. We were to assist, mentor and train the ANA while ensuring that we all returned home safely with a good attitude toward the Guard and future deployments. The last two were a bridge too far but that was hard to see then.

Force Protection, part of the monolithic Risk Management system would be the primary concern for all subordinate leaders, and personnel accountability was emphasized. OPSEC and personal security along with physical fitness routines and leisure activities were all part of the CO's overall concern for our wellbeing and he meant to hold the subordinate leaders accountable for assessment reports regarding these things. Once again, the last two would be difficult and eventually swallowed up by the tempo of our assignment.

He wanted the junior officers and the NCOs to be able to make decisions without fear of their seniors second guessing them, stating that decentralized execution would overcome the volatility, uncertainty, complexity and ambiguousness that is prevalent in this environment. Those conditions would certainly be there but I'm afraid that our top down command habits were hard to break. Counseling was required monthly but if it was done it was a paper drill in the face of long days of actual work. He had a realistic outlook on the task we were given which has been described as "Eating soup with a knife," and urged patience toward our Afghan counterparts.

His words on communication were interesting and had he the time he could have trained the rest of us how to communicate with him. Since we didn't use what time we had for that, eventually most of the younger officers and the NCOs kept their mouths shut when they encountered him. His emphasis on communication was extended to the daily command and staff meetings which were not to exceed 30 minutes—a lofty goal.

The last item on the CO's command philosophy was the exhortation to maintain a sense of humor. Since he didn't appear to have one, this was something that took on a life of its own, as we did like soldiers have always done when facing the unknown, and turned all of the commander's philosophy into our own inside joke.

The Physical Exam

We had to get physicals and a dental checkup. I got some new glasses. I found out that my hearing was getting worse but not bad enough to keep me from going. They didn't seem to mind if a person was overweight and there was no PT test to pass before you could deploy. Unless a person had a serious medical problem they were going.

Pre-Deployment Training

After getting our gear and being certified healthy, we began our pre-deployment training. The Army had devised a system over the years that now utilized Army Reserve units to certify National Guard units as ready for overseas deployment. This certification system was started in WWII and minutely adjusted for the Korean War. They didn't need it for Vietnam and it might have gotten a few tweaks for the Gulf War after which it lay idle until 911.

There was a long list of tasks, mission essential tasks, that someone came up with and we had to check the blocks on each one until we were 'certified' as combat ready. Most of the tasks were common soldier skills that everyone should have been familiar with if they had spent any time in a combat arms unit; but whether we knew how to do these thing or not, they were on the list so we had a class. Usually in the Army these tasks were taught using the performance oriented training method of Tasks, Conditions, and Standards with performance evaluations and retesting until you got it right. This was more of an orientation and it didn't matter if you were clue free at the end of the class; when the training schedule indicated that the class was over it was time to move on.

In a perfect world the unit we were replacing would have developed a list of tasks based on what they were actually doing. We would have had relevant training on equipment being utilized by the unit we were replacing. We would have discussed these tasks on the video teleconferences (VTC) which were capable of facilitating face to face communications. I went to a couple of them and found that they were next to useless for passing on useful information. Since they were usually presided over by general officers and colonels, the senior officers felt it was incumbent upon them to make brilliant statements about various subjects that had something to do with the mission. The junior officers were intimidated by their presence and wouldn't dare ask a relevant question for fear of being accused of being stupid. The VTCs were always scheduled for U.S. time so that meant that the attendees in Afghanistan, which was almost half-way around the world, had to be up at some ungodly hour to accommodate the warriors back in the states. This was a common occurrence at the Pentagon and CENTCOM so no one at our level gave any thought to taking care of the warfighters' sleep schedule.

We had some first aid medical training and applied field dressings to one another's wounds. They tried to explain the SINCGARS radio system to us but since we never saw any of that equipment after we left, we agonized over our ineptitude for nothing. They gave us a class on the Plugger, a military GPS, and we all walked around a large parking lot trying to find a waypoint. We never saw those again either. Many of us had small, cell phone sized GPSs of our own but no one had mapping software for Afghanistan.

Have Gun—Will Travel

We were issued weapons at Fort Carson. The RTI didn't have any and apparently the Army was short of modern small arms that fall. Most units were deploying with the M4 Carbine, the latest version of the M-16. The M4 usually had special optical sights that could be used in the dark. The next upgrade might be a small flashlight mounted on the barrel. Then came an infrared laser projector that would let you put a pulsing light beam on your target at night that could only be seen with night vision goggles. Finally, there was a front hand grip known as the 'gangster grip' and a special sling for carrying your high speed weapon of war while you took pictures.

While everyone in the 45th Brigade would have a variation of the M4 in Afghanistan, we were issued perhaps ten of the basic M4 without any extras. The majority of the weapons issued was the M16A2. It had an improved plastic stock and handguard, and instead of full-auto rock and roll, it had a three round burst capability. It was a full sized rifle and was not designed for shooting from small vehicles. Pistols were also in short supply. I had been carrying one for the last 18 years so of course I was issued a rifle. It was a pain to carry and I was determined to somehow get one of the pistols that had been distributed haphazardly to various members of the unit.

We had weapons training which consisted of a class on instinctive shooting taught to us by the Reservists. They were sure that they knew everything that we needed to know about operations in Afghanistan, even though none of them had been there nor would they be going. I caused some trouble when I insisted that we change the targets so we could get down-range feedback rather than just burn up all the ammo so that they wouldn't have to turn it in at the end of the day. During a lively discussion with the CO, who wanted to know why were weren't heading back to the barracks, I told him we needed to get clean targets if we were going to fire more ammo, whereby he ended my marksmanship doctrinal demands by telling me "at ease" and told the troops to fire up the rest of the bullets so we could go back and eat supper.

The next day we were required to qualify with our rifles, they said, and I was one of a few that managed to qualify the first time. I got to go back to the barracks and I never heard about how the others fared the second time around. I assumed that they all greatly improved their marksmanship because we weren't going back to the range again and everyone was ready for the next class.

Driving Miss Doozy

The next two days we had driver training. They bused us out to a small army airfield where we met a group of young men in civilian clothes who taught us how to drive in Afghanistan. They were a team of SF soldiers from the Colorado National Guard who had recently returned from Afghanistan and their team sergeant hired them all to work on contract, teaching driver training and tactics to units like ours.

The first day was driving skills. We learned how to do the 'bootlegger turn' where you were speeding down the road and suddenly decided that you wanted to go the other way. You turned the wheel slightly while pulling up hard on the emergency brake. As you slid around on your rear tires you released the emergency brake and pushed down hard on the accelerator and found yourself going the other way in a cloud of dust. That was fun.

The next thing was ramming through a road block made by vehicles. After that we chased after another car driven by our comrades and steered into the rear side of their vehicle causing them to spin out right in front of us. By the end of the day the cars that they had brought with them looked like the last hour of destruction derby, but we all agreed that it was the best, or at least the most fun training we had had so far.

Drive-By Training

The next day we were bussed out into the foothills where there was a range that we could use for firing from our vehicles. After some classes on tactics we loaded some rifle magazines with live ammo, divided up into three vehicle groups and drove around the range reacting to ambushes and rescuing a disabled vehicle team under fire. We weren't the Special Forces but no one shot the guy next to him so I figured if we really had to do this we would just have to do the best we could like National Guardsmen have done in all our wars.

Thanksgiving

We were scheduled to leave at the end of December and Thanksgiving was approaching. Fort Carson was shutting down for the holidays so I suggested to the CO that we move the whole unit out to the field where we could conduct small unit tactics and night patrols. He decided to let everyone go home for a week instead.

I didn't go because I had already been through one painful leave taking so I decided to stay at Fort Carson and continue my anxiety spending spree. I found a company on the internet that was selling satellite telephones and had them send me one with a hundred minutes. These were truly a modern marvel. You could stand in the open anywhere on earth and talk to a land line or cell phone anywhere else on earth. Since there wasn't any contingency planning being done by the staff, I decided that I would do my own and the first thing you need in most situations is a way to call for help. I had the phone Fed-Ex'd to me and got it before we left.

There was a family of believers in Colorado Springs and they took me with them to a meeting of the church in Denver. I felt at home there and asked the saints to pray for us in Afghanistan. I went with the family to a home meeting in Colorado Springs one time before we left.

The 136th RTI, Camp Mabry, Austin, TX
(author photo)

Fort Carson, Colorado
(author photo)

Offensive Driver Training
(author photo)

Offensive Driving Training
(author photo)

Quick Fire Training
(author photo)

No one serving as a soldier entangles himself with affairs of this life, that he may please the one who enlisted him.
2 Timothy 2:4

Chapter Two: December 2003

Travel Plans

After the long break, finally, we checked all the blocks on the list and were ready to go. One of our officers Major B, had been busy making arrangements for our transportation. He managed to get a C-17 that would take us to Kabul instead of Bagram Airbase, saving us a long truck ride. The only problem was that we weren't taking enough weight to justify the flight so the major ordered several tons of bottled water and we were good to go.

Turn On The Fire Hose

About two days before we left, we met a representative of the unit we were replacing, an LTC from the Vermont National Guard. He began to try and explain all of the things they were involved with in the performance of the mission. There were so many different aspects that eventually it all ran together and we entered into a phase of the operation that was called 'drinking from a fire hose.'

I could tell that they had faced some challenges but while I didn't think he was being totally candid about what was going on there, I had no idea. I did find out that I would be supervising the interpreters as part of my S2 duties. I also found out that there were some secret documents that his unit's S2 had stored in a safe in the headquarters before he went home, earlier than the rest, for some unexplained reason, and when we found the key I could see what they were. As far as I know they are still in that safe. The last thing he told us was that we would have a right seat/left seat ride where we would work with the one we were replacing for a week before they went home.

The 'Heroes' Depart

The night before we left there was a going away ceremony for us and another unit that was also going to Afghanistan. Fort Carson's

Commanding General, the Chaplain, and the Command Sergeant Major all told us that we were heroes. This was something that the modern army had started putting out for a war that would have few opportunities for anyone to be a hero in the classic sense so they tried to convince us we were all heroes. The band played on as we filed out for one more night at Fort Carson before we flew.

Essential Tasks

The next morning we were issued our weapons but no ammunition. We loaded our gear and got on the bus. At the airbase, while we were waiting to get on the plane, we had a class on weapons clearing. There had been so many negligent discharges at the clearing barrels placed just inside every base in Afghanistan that the command emphasis on weapons training for soldiers going to war was the correct method for unloading a rifle or pistol at a barrel filled with sand. We all had to stand in line and demonstrate our ability to remove a magazine from our rifle or pistol or in some cases both, to the CO and the CSM.

We were still waiting to get on the plane when the CO ordered me to give a particular OPSEC briefing to the unit. The main point was that all the members of our coalition were actively collecting information from their coalition partners. I directed our unit to safeguard all information from anyone who didn't have a need to know and left it to them to figure out what that meant since we would be working closely with the Brits and the French.

All Aboard

After a long wait the plane was ready for us to board. We walked across the tarmac with our weapons and a carry-on bag. When we got inside we grabbed a seat. I got one in the front of the aircraft and was happy with my choice when later, I heard that the people sitting in the back were freezing due to some heater malfunction.

We finally got everything tied down and the plane began rolling. There were no windows near the seats so I closed my eyes and enjoyed the sensation of hurtling down the runway until the plane began to lift off the ground. I could feel the change in vibration and then I could hear the wheels being drawn into the body of the plane. The loadmaster dimmed the lights and we flew through the night sky toward Germany.

Sometime during the flight we did an aerial refueling over Greenland but it was so smooth that only a few people even knew it was happening. I woke up every so often and tried to get my body into a more comfortable arrangement before falling back into sleep.

The next morning we were told to fasten our seat belts and shortly after that we slid into a smooth touchdown at Rhine Main Air Base in Germany. We were put on a bus and driven to some large tents that were being used for transit quarters. There were many cots so I took one at the far end of the tent and took a nap. The CSM woke us all and announced that we would be here for two days while some repairs were done on our plane. I spent the next two days eating pizza and buying whatever cold weather gear they had in the Base PX.

All Aboard—Again

Two days before Christmas, after dark we got back on the plane and flew toward Afghanistan. I got another seat close to the front of the airplane where it was warm. There was a window near my seat and I planned to take some video through it when it got light. The CO found me in my seat and told me that there would be another mid-air refueling and that he had gotten permission for me to go into the cockpit and videotape it. I rushed up there and shot video as the crew maneuvered our aircraft while moving at about 300 miles per hour in the dark over the Black Sea in Russia.

The pilot got the nose of our plane under the tail of the KC-135 tanker aircraft and the boom operator in the tanker was able to move the fuel nozzle into a small hole in the top of our fuselage behind the cockpit, and fill us up with enough gas to get us to our destination. I thanked the pilots and commended them on their skill which they, being flyboys, acknowledged as being something that they did every day.

When I got back to my seat I took a nap for a few hours and when I woke up I could see that it was daylight outside. I stood up at the small window near my seat and took some video of the snow covered mountains. There was no sign of people in the vast white landscape. No houses, roads or any sign of man's existence.

This was Central Asia, land of the Great Game[68] and the natural barrier between Russia and the Indian sub-continent. I wondered how long we would last if the plane went down here and we survived the crash; but the plane flew on with no sign of weakness and soon we were told by the loadmaster that we were an hour out. I went back up to the cockpit and asked for permission to videotape the landing but the pilot told me that we were going in under combat rules so I would have to return to my seat. I took more video out the window until we were about to land and felt through the seat of my pants, the first contact of our wheels with the runway of the Kabul International Airport.

Kabul, Afghanistan

We got off the plane and the first thing I saw was a group of French soldiers, including a pretty girl, with automatic rifles, standing nearby on guard duty. It was cold and overcast. There were also some U.S. soldiers with trucks nearby and one of them was carrying ammunition which was issued to us for our use if we were attacked on our way to our base about 10 miles away. We were in the war. We loaded our magazines and then climbed into the back of the trucks with our weapons and carry-on bags and tried to get comfortable while facing outward.

The ammunition they issued us was enough for a peacetime basic load for our rifles. The pistol ammunition was in short supply but I was sure that we would get more when we got to Camp Phoenix. That was wishful thinking and eventually we had to trade coffee and other supplies for ammo with units that were passing through. This lack of ammunition was but another indicator that the picture painted by Colonel Knight was BS.

The trucks followed a Humvee and we drove through the capitol city of Afghanistan. All the people looked poor and cold. Many of the men wore turbans and baggy pants. Most of the men wore sandals in the dirty snow and slush on the streets. We saw the famous 'burka babes,' as the Rhode Islanders called them, walking along behind their men. The kids were running everywhere and seemed to be happy in their circumstances as kids always seem to be. I tried waving to anyone who caught my eye and was pleased to notice that 90 percent of them waved back and smiled. This was my method of determining the mood of the people in Kabul with regard to our presence.

The air quality was terrible. It was winter and there was little fuel for the poor people to burn to keep warm. They were burning things like the plastic water bottles they found in our dumps and other environmentally unsound material such as feces. The air over Kabul reminded me of the air in LA on a no wind day. Kabul was surrounded by mountains which kept the bad air from going someplace like Pakistan to the east. I guess the air also carried germs because within days of arriving I came down with a horrendous sore throat. When I couldn't swallow I reported to sick call. They gave me a shot of a strong anti-biotic and two days later I returned to work, the only time I was sick in Afghanistan.

Camp Phoenix

We finally arrived at our new home, Camp Phoenix[69] on the east side of Kabul, and the north side of Jalalabad Road. There was a man with a big black beard who stood at the entrance to the camp. He saluted each truck as it passed him. We found out later that he was known as 'Rambo' and while that seemed like a pretty ridiculous nickname, later events would show that it was well deserved.

The camp was situated in an old Russian transportation facility which had some large warehouses filled with wrecked machinery that we passed after we entered the gate. After we passed the warehouses the truck turned right and came to a stop where we had our first dance doing the clearing barrel polka. There were some serious faces as the troops carefully followed the procedures and we managed to get everybody through without any loud noises. The holes in the barrels stood as mute reminders of the hazards lurking at one of the most dangerous places in Afghanistan at that time. We wouldn't always be so lucky.

The trucks took us down long rows of tents which we could see over the top of a wall that was constructed with HESCO barriers, a pre-fabricated wire mesh and fabric cage filled with dirt and small rocks. These HESCO barriers surrounded the camp with two rows, the outer wall and the inner wall which were separated by about 50 feet between them.

The space between the HESCOs was used for the large generators that provided electricity for the tents and offices in the wooden huts in the center of the camp. The vehicles were also parked in this space and most of them were Toyota SUVs and small white Ford Ranger pickup trucks with crew cabs.

We climbed off the trucks and a major from Rhode Island did a drop test of my laptop that I had stored in my backpack. When I called his attention to what I considered a pretty unthinking act he just shrugged as if to say, 'it happens.' I formed the opinion that 'it' probably happened a lot in his life and the next six months would confirm my suspicions. Fortunately I had a Dell which was from Texas and it survived the 5 foot drop and worked perfectly for my whole deployment.

We walked through a gap in the HESCO barrier and found our duffel bags which were stored on pallets. The CSM had billet assignments and I found my tent which was next to the supply room about 10 feet away from where I was standing. The tents were surrounded by walls of sandbags; I was glad to have missed that work detail. Inside there were rows of bunks, five to a side. I took an empty one on the left side, close to the door and began to unpack my sleeping gear. There was a place to hang my uniforms and a small storage area for my underwear and socks.

Camp Phoenix was situated very close to the flight path for the Kabul Airport. Every day, all day, a steady stream of civilian and military aircraft flew close-by overhead. The Russian transports, the largest planes in the world seemed to just float over the camp. Occasionally a military aircraft on approach would fire off their anti-heat-seeking missile flares, which would start fires in the Afghan villages nearby. This didn't do any good for the hearts and minds effort.

The 45th Infantry Brigade

The coalition of the willing that was committed to the rebuilding of the Afghan National Army was led by the 45th Infantry Brigade of the Oklahoma National Guard. Since they shared a border with Texas I was sure we would be welcomed by our neighbors to the north. I couldn't have been more wrong. Oklahoma and Texas had been nursing a grudge for a long time. I couldn't remember reading anything about it but I assumed it had something to do with the location of the borderline.

Once again I couldn't have been more wrong. The long standing grudge between Oklahoma and Texas had nothing to do with territory, unless you considered a football field as territory. Oklahoma had kicked Texas' ass in the Cotton Bowl that year and they weren't going to let the Texans forget it.

I was a high school drop-out and a surfer who had moved to Texas from California. Being uneducated and not interested in football, I was totally unaware of the issue that had consumed the fine people of these two states for all these years and which threatened to derail our best efforts in the great war on terror.

The 45th Brigade CSM

After an hour we were directed to go to a building near the headquarters where we would be briefed by the CSM of the 45th Brigade. After a ten minute wait he showed up and acknowledged our presence; he told us that Camp Phoenix was a wonderful place and he hoped we would be happy here. He was quite excited to tell us that a contract had been signed with a laundry company and that soon, we would be able to have starched uniforms. If he expected shouts of joy, what he got was radio silence, so he moved on.

Then he informed us that if we wanted a medal at the end of our tour we had to make sure we didn't have a negligent discharge at the clearing barrel. His power point briefing followed this line of CSM reasoning and I was glad that I had been reassigned as the S2, a position that also had the potential for extremely stupid proclamations but I felt that I had enough experience to be able to mitigate those tendencies with some reality. The CSM droned on about his housekeeping concerns and finally asked if there were any questions.

No one seemed to need any more from him but I was interested in a certain aspect of the camp layout that he hadn't covered and raised my hand. "Sergeant Major, can you show us on the map of the camp where the bunkers are for mortar and rocket attacks?" There were no bunkers marked on his map, just the DFAC, PX, the gym and the laundry. He seemed a little annoyed with this question and said that we would have to explore the camp and find them on our own.

That told me two things: one, he was an admin guy and two, things were peaceful in this part of Kabul. If there were any incoming, those bunkers and the locations would have been the first item of concern in his briefing. After the briefing, I did explore the camp and found some bunkers about 100 yards from my tent and decided that if we did get incoming, I would stay in my tent behind the sandbagged walls rather than risk running through the camp and getting shot by someone.

The CSM's hope for our happiness at Camp Phoenix didn't extend to the members of his own unit who were caught sleeping on guard in the towers around the camp. They were forced to wear signs around their necks stating that they had fallen asleep on guard. The German army used to hang signs around the necks of deserters at the end of WWII along with a rope which they used to lift the deserters by the neck to the tops of the lamp posts as a warning to others. I'm sure such a practice would have substantially reduced the number of sleeping on guard incidents but this was the National Guard.

The Combined Joint Task Force Phoenix

We ate dinner at the DFAC or dining facility. It was in the middle of the camp near the headquarters. As I stood in line I could see the headquarters building and the collection of flag poles that were flying the flags of the many nations that were part of the Combined Joint Task Force Phoenix.[70] I recognized the British Union Jack, the Canadian Maple Leaf, and the Southern Cross of the Australian and New Zealand flags. The French tricolor was there but the rest of the flags I didn't recognize; in the weeks to come I would find out that there were troops from Romania, Bulgaria, Germany, and Mongolia.

Task Force Phoenix was originally manned by the 10th Mountain Division. They oversaw the mission of rebuilding the Afghan National Army. Task Force Phoenix II was commanded by the 45th Brigade of the Oklahoma National Guard and from then on Task Force Phoenix would be a National Guard mission. The goal of TF Phoenix II was to build the Central (Kabul) Corps, a multi-brigade unit of the ANA. This effort would eventually expand into five more Corps around the country.[71]

The actual training of the Afghans was conducted initially by the British Army who trained the 1st Kandak, an ANA battalion. The 3rd Special Forces Group trained the next Kandaks and began building up the Kabul Military Training Center (KMTC). Soon the French Army joined the Task Force along with the other nations represented by the flags in front of the HQ.[72] As usual the U.S. was paying for everything which meant that the other nations' soldiers had to eat our food, live in our tents, and conform to our rules.

KMTC

The Afghan National Army (ANA) was being reborn at KMTC. The Afghan training cadre had been developed and we were to provide 'assistance' in the form of logistical support and 'mentoring,' a popular word in the military at that time. Everybody was supposed to be 'mentoring' someone but as yet most of us had not been told what that meant so we gave advice or tried to do what needed to be done ourselves.

The training plan was one of the craziest ideas I had ever heard of. Three different nations would take responsibility for establishing a fourth nation's army. The French ended up running the officer training. The British ran the sergeants' training, and the Americans ran the basic training of all the soldiers. How and if the training doctrine was synchronized by the four elements with three languages is something that I never found out. We inherited a program that was in place and it was our job to keep it going regardless of the apparent inconsistencies in logic.

The British Army

The British Army trained the Afghan sergeants. Since the sergeants are the backbone of the British and American armies, this seemed like a good fit. The British contingent at KMTC came from the 2nd Battalion, The Royal Gurkha Rifles. The British fought them during their 'business' expansion in India and somehow convinced them in 1815 to join the British Army where they built a fearsome reputation on many battlefields. Since the Gurkhas came from this part of the world many of them spoke Urdu or Hindi and were able to communicate with some of the Afghans who learned those languages both in the refugee camps of Pakistan and watching the movies from Bollywood.

The British officers had to learn Gurkhali, the language of the Gurkha soldiers who were recruited in Nepal. The Gurkhas are considered an elite unit and there is a long tradition of Gurkha officers rising to the highest ranks of the British Army, Field Marshal William

Slim of Burma fame among them. An excellent memoir by John Masters, "A Bugle and a Tiger," gives much detail about life in a Gurkha battalion in the 1930s. His story takes place on the Northwest Frontier of India where he fought the Pashtun tribesmen we are fighting today in Afghanistan.[73]

There were about 30 Gurkha NCOs and three British Gurkha officers. The Gurkhas were easily the most experienced, professional and most fit soldiers in the task force. They had a morning run led by their officers every day.[74] They all carried a distinctive knife, the 'Khukuri,' everywhere they went. Legend had it that if a Gurkha drew his knife it had to 'taste' blood and if he didn't cut an enemy he would cut himself before returning it to its scabbard. If it's true they suspended the practice while working with us because they were besieged by the Americans at Camp Phoenix to show them the blade. They had 'Curry Night' once a week at Camp Souter in Kabul and a fortunate few were invited.

Like I said the Brits were the most professional etc. They also had a great sense of humor. They took great delight in pulling at the tail feathers of the Thunderbird, the symbol of the 45th Brigade, with subtle digs at the Yanks in their Power Point slides at the General's daily battle update briefing, the 5 o'clock follies now known as the BUB.[75]

The French Army

The French contingent appeared to come from different units of the French Army. Because of the language barrier we had even less contact with them than we did with the Gurkhas. During the buildup to the U.S. led invasion of Iraq in the spring of 2003, in which the French and Germans declined to participate, there was a storm of criticism mainly aimed at the French. There were quite a few very insulting articles on the internet that made our job of interfacing with them even more difficult.

Fortunately we had a sergeant from East Texas who could speak Louisiana French and he became the liaison with our French counterparts. Our officers were invited to some wine and cheese parties. Since the French NCOs were required to carry their officers' baggage, we sergeants would have been invited only if we would help them in pouring the wine and cutting the cheese.

The Mile High Cafe

The inside of the DFAC had a sink where everyone was required to wash their hands. Then we had to sign a headcount sheet. There was a young Afghan man named Matt assisting with the sign-in and as time went by he would become the most significant person of my deployment. The DFAC was run by Kellogg, Brown, and Root (KBR) my old friends from Vietnam. Brown and Root had some big contracts in that war and now, joined by a new friend Kellogg, they were at it again.

No more KP—the civilian contractors did all the menial work that had been done by soldiers in the 'old army.' KBR cooked and served our food, maintained our vehicles and generators, did our laundry and ran our morale center and gym. They hired local help and that was how Matt came to be there helping us to sign in. After signing in I went through the food line and found a table. Everyone carried their weapons at Camp Phoenix and it wasn't unusual to see a machine gun resting on its bipods on the floor next to its owner's table. There were a lot of snacks and ice cream available and I could see that quite a few soldiers were starting to look fat.

The RIP Begins

After dinner I went back to my tent. We had been alerted by our CSM to be ready to leave in the morning for a drive to KMTC where we would begin to take over our duties from our predecessors from Vermont. This was called a RIP or 'relief in place.' In the infantry you would leave your machine gun tripods in position and the relieving unit would give you theirs. In this case the Vermont unit would leave us with all the problems that they hadn't been able to solve along with a few they created and we would give them our best wishes for a safe trip home.

Jalalabad Road Running

The next morning I went out to the motor pool and found LTC Randy Davis and some of the Vermont troops. We loaded and locked our weapons. Our counterparts drove and we watched for ambushes and suicide bombers as they sped down Jalalabad Road, a four lane divided highway. There was a lot of traffic on the road. Taxis, vans full of people, buses, large trucks, small trucks, military vehicles from the 10 or so nations that were patrolling Kabul, and privately owned sedans and SUVs, all jockeyed for position in the great Kabul-Jalalabad Road race. The slower traffic like donkey carts and camel trains stayed to the right, generally on the shoulder but sometimes on the edge of the right lane. In the middle of it all were the 'Jingle' trucks.

I'm not sure why they are called 'Jingle' trucks because I never heard them 'jingle. I was diagnosed with hearing disability so that may explain it. My eyes worked OK and it would be hard for anyone but a blind man to miss seeing these rolling pageants of Central Asian artwork. There was everything from tigers' faces to green pastoral scenes, both of which were rare in Afghanistan. They were usually loaded with much more cargo and people than could be safely carried by our standards, but the Afghans were on the move and these colorful trucks could be seen everywhere.[76]

We drove in tactical zig zag formations with the lead vehicle in the left lane and the second vehicle in the right lane and so forth. We were told not to let anyone pass us, so the rear vehicle would swerve from left to right whenever someone tried. Most of the local Afghan people knew we were a military convoy and would remain behind us but some would give in to road rage and speed past us, driving down the unpaved divider strip, about 3 lanes wide in the center of the highway. This would raise clouds of dust as they raced ahead.

The Kabul Military Training Center

KMTC is on the north side of Jalalabad Road, across from Camp Warehouse where the ISAF troops lived. The International Security Assistance Force kept the peace in Kabul and consisted of soldiers from Germany, Norway, Italy, and Great Britain.[77] We turned left and drove past an unoccupied guard post with a big picture of Ahmad Shah Massoud on its front and down a road which took us to a side entrance to KMTC. The Afghan soldiers waved us through and we drove past some bomb damaged buildings and some new construction until we came to a HESCO barrier and a gate that blocked entrance to the Alamo.

The Alamo

The gate was manned by American MPs who looked us over and opened the gate. These MPs were from an Army Reserve unit that had been called to active duty right after 911 and had spent a year guarding the Pentagon before being sent to Afghanistan where they were split up and now were guarding various bases. They were not happy in their work and the Vermont soldiers were not happy with them. I began to see a trend of unhappiness that had infected the troops that we were replacing and hoped that we would not catch the same sickness but I'm afraid that for some it was inevitable.

The 'Alamo' was what they called the small compound that was occupied by the American, British and French soldiers that worked at KMTC. It was so named by the first SF team to come here. They were surrounded by thousands of Afghans being trained here and knowing history, they fortified this compound as a last ditch fallback position if the Afghans were to turn on them.[78] Its name may have been the inspiration for Colonel Knight's naming the TF after the Rhode Island captain who died at the Texas Alamo.

It was surrounded by HESCO barriers and it had several guard towers manned by MPs. There was an old U shaped building in the middle of the compound and that was where our offices were. Tactically, it couldn't have been in a worse place. Just on the other side of the HESCO at the open end of the U was a tall hill that overlooked the compound. There was a pump house and water tank that was used by KMTC but the hill was outside the perimeter. It was a perfect place to put a couple of machine guns or a SPG-9 if the enemy ever wanted to attack the Alamo. I voiced my concerns to the CO the first time I saw it and he discussed it with the Vermont CO who told him that they had requested a security detachment from Camp Phoenix to occupy the hill but so far there were no troops available so they learned to live with it.

The MPs who provided security for the Alamo had recognized the threat and kept a Humvee in the courtyard with a Mark 19 grenade launcher aimed at the hilltop. This was a comfort to some of the people but I was sure that if we came under attack from a determined group of mujahideen, they would take out that Humvee with the first volley and own the courtyard.

KMTC

KMTC was formerly the Military Academy of the Afghanistan Army. The headquarters was in a five story building known to us as 'The Five Story.' Now KMTC was the training base for the 'new' Afghan National Army (ANA). There were about 5000 Afghan soldiers here going through basic training. There were a few hundred more soldiers going through a non-commissioned officer course that took basic training graduates who showed potential and trained them to be sergeants. There were more soldiers who were going through an officer training course that would graduate them as lieutenants. Our job would be to 'assist' the Afghan staff with training advice and logistical support. The 'advice' was accepted because the main thing they needed was logistical support and since the Transitional Authority didn't have any resources, the ANA was totally dependent on us.

Our people would soon be arranging construction contracts costing millions of dollars and overseeing the building projects. They would be arranging the feeding and clothing of thousands of soldiers that were passing through KMTC every day. Most of our people had never done anything like this in the military and I was amazed at how quickly most of them figured how to do their jobs and got to work.

This is one of the National Guard's greatest strengths, the ability to adapt quickly to new situations. This is because of the diverse life experiences of its members who were part-time soldiers with jobs or businesses that had taught them much more about life than the average regular army soldier of the same rank.

The Eyes Are The Window To The Soul

The Vermont soldiers had taken on the fashion statement of the SF. Wrap-around sunglasses were the most obvious addition to the Army uniform. The Army issued all of us a pair of Wiley-X sunglasses and almost everyone was wearing them. Though I hadn't heard about a health hazard caused by the unusual effects of the sun in Afghanistan, I had toyed with the idea of wearing them back at Fort Carson, being as fashion conscious as anyone; but I was unable to get prescription lenses for them so I gave up the idea and I'm glad I did.

There's nothing worse when trying to communicate than being unable to see the eyes of the person you are talking to. The eyes are said to be the window of the soul. In my police experience it was always one of the places I checked after ensuring that the suspect's hands were not a threat. The Americans came to Afghanistan wearing sunglasses and that alone has to be one of the contributing factors to our slow progress in that country. I never wore sunglasses while I was there and I think it was a help in my dealings with the many Afghans I worked with.

The Son Of The King

The LTC led us to the building entrance in the courtyard of the U shaped building. When we got inside he introduced me to Bill, a short heavyset master sergeant, who I was replacing. He bore an uncanny resemblance to the last king of Afghanistan. He was totally bald, with a hawk beak-like nose and a conspiratorial smile hovering around his mouth. He had a big laugh and he didn't appear to take his situation very seriously. He had been supervising the interpreters and he was going to stay in Afghanistan after the rest of his unit returned to Vermont.

Bill was an energetic man and he had been very busy here; actually he was a whirling dervish. He was juggling a number of different projects that he had come up with on his own and he wanted me to carry them on. These projects included translating military manuals into Dari, the language used by the Afghan military; establishing and running schools for workers at KMTC and Pul-i-Charki and providing the funding for materials and wages for the teachers. He was the middle man for a leather shoulder-holster business for one of the interpreters and he had a number of other initiatives that would require much of my attention in the months to come.

He began to brief me on all these activities as soon as we met and I was soon overwhelmed by the scope and range of his projects. He was spending three or four hundred dollars a month on his education projects, the money coming out of his own pocket and from donations from KBR workers at Camp Phoenix and other people around the Kabul area.

He was making deals with people for equipment and resources and every day when we got together he would reveal a new project of his and begin to tell me how I needed to care for it and what I could do to make it better. Little by little I was able to organize all these various activities in my mind and on paper and I began to evaluate the situation using the Special Ops doctrine that I had been immersed in back in SOD.

To sustain some of Bill's projects for the long term was not going to be possible for me and after a few days I told Bill. He was disappointed but when I explained my doctrinal basis for my decisions he was able to see that I had a reasonable approach to the problem and we agreed to disagree about certain aspects of his efforts.

C-17 high over Central Asia
(author photo))

Afghanistan
(author photo)

Headquarters, Coalition Joint Task Force Phoenix
(author photo)

Camp Phoenix, December 2003
(author photo)

(author photo)

'It happens'
(TAG photo)

Suffer evil with me as a good soldier of Christ Jesus.
2 Timothy 2:3

Chapter Three: January 2004

The Daily Staff Meeting

We began having staff meetings every morning. The CO let us know that he wanted the bottom line up front when we made our presentations. Each staff member was required to give their Power Point slides to Major Shawn the night before. He and his crew would put them together and we had to be ready to go at 0800. I was always working on these slides to the last minute and usually gave my slides to Major Shawn about ten minutes before the meeting.

Some of the officers had been playing the Power Point games for a long time and they were able to dazzle most of us with their clever graphics and transitions. The CO let us know that he was only interested in the meat of the matter and advised us to leave the skin and the feathers at home. He hammered various staff members because he only wanted the basic facts, not a long sea story.

45th Brigade Intelligence Section

The CO told me that he wanted an intel update available to everyone before we left Camp Phoenix for KMTC each morning. This would require me to get the latest updates from the S2 of the 45th Brigade several hours before our scheduled departure. I walked up to the 45th Brigade headquarters and found the S2 offices. There was a major sitting at a desk with a sign proclaiming 'Dar al Harb' a Koranic term meaning the House of War.

'Dar al Harb' was in opposition to 'Dar al Islam' or House of Islam. The Muslim doctrine being followed by the Taliban explained that the Afghans had defeated the Russian infidels, the Taliban had purified the Jihad, and after Mullah Omar had wrapped himself in the 'Prophet's' cloak in Kandahar and declared himself "Commander of the Faithful," Afghanistan had returned to 'Dar al Islam.' Now the infidels were back and we were all in the 'House of War.'

Behind him on a bookshelf were a number of books about Afghanistan, some I had read and some that I would like to read. I interrupted his reading of a document and introduced myself as the TAG II S2. I told him that my CO had directed me to prepare an intel briefing each morning and asked if the S2 shop could give me what threat information they had every morning about 0400. I heard another voice come out of the office to my left. Another major came to the door and asked me my business. He was the actual S2 and the major I was speaking to was his deputy.

This major had a small mustache and he didn't look friendly. In this case, looks were not deceiving and he proved to be an adversary from the start when he told me that we didn't need any intel. "That's a slippery slope, Bobby," said the major sitting at the desk. After they discussed the matter while I was standing there, the S2 begrudgingly agreed to allow me to come to the TOC at 0400 and get any updates. The major who intervened for me had the name Mancino on his uniform and as I left I wondered if he was any relation to BG Mancino, the 45th Brigade commander. As it turned out, he was the general's son.

My Daily Intel Summary

The next morning I was up at 0330 and I walked, crunching ice and snow, on a path through the rows of tents towards the headquarters. I shaved with my battery powered razor as I walked along, just like I had in Vietnam 36 years before. When I reached the headquarters I walked through the quiet rooms where bleary eyed night shift people struggled to stay awake during the lowest point in their circadian rhythms.

The S2 section was manned by a young second lieutenant who was friendly and gave me everything they had as far as intelligence about enemy threats and plans which was nothing. They did have a lot of reports about terrorists in yellow and white vehicles who were ready to attack the infidels at any moment. All the taxis in Kabul were yellow and white. Some of these reports gave the license plate numbers of these attack vehicles. This might have been helpful if we could read Arabic script but we hadn't yet mastered that skill and the thought of our guys trying to learn as they careened down Jalalabad Road seemed to work against our risk management procedures.

Nevertheless, when I reported to the CO the state of intel at the 45th, he directed me to summarize these reports and produce a briefing paper with vehicle descriptions and license plate information, which I did. Every morning I got the intel dump on a 'flash' or 'thumb' drive, took it back to my tent where I quietly cut and pasted the information with the weather forecast onto a one page document in my laptop and printed it. I took this over to the supply room and made enough copies so that each vehicle driver could get one. The troops read these things for the first few days and I got a few questions about the veracity of the information. I told them all the same thing; this was hot info from the brigade S2 shop. I didn't want to undermine the CO's strategy of keeping everyone alert for threats as they drove to work and back because in Afghanistan you never knew.

The Vermont RTI RIP Continues

The guys from Vermont were ready to go. Most of the unit had left before we got there; so much for the 'left seat/right seat ride' that we had been assured of at Fort Carson. The TAG commander and his senior officers and some of his senior NCOs were still there and it was easy to see that they had had a rough tour and were ready to go home. Some of them were not speaking to one another and eventually I found out that a number of them had lost the confidence of their comrades and though they were trying to conceal their dirty laundry from us, you could feel the tension as they tried to explain the details of our mission to us.

Drinking from a fire hose was the term used by Major Shawn and it fit the situation perfectly. The sensory overload caused most of us to be unable to organize all the information that was being given to us. I continued trying to reconcile what we were seeing on the ground, what we were being told by the people we were replacing, and what we were told by Colonel Knight before we left. This created some dissonance in my brain and it took longer to assimilate the information I was getting from the various sources in my sphere of operation.

The Interpreters

One of the first things Bill did was to introduce me to JS, the lead interpreter at KMTC. He was dark haired, clean shaven, with a handsome face. He had a very pleasant demeanor and was God's provision for me in my job. Without JS I would have been lost and he became my closest friend in Afghanistan. JS was from Kabul and he came from a Pashtun family that was originally from Nangarhar Province. His father had been an official in the DRA government and had retired after the Taliban captured Kabul.

JS had attended the Military High School and had hoped to be a pilot in the Air Force. When the U.S. came to Afghanistan, he applied for an interpreter position with the Special Forces who had begun training the Afghan Army at KMTC. He had worked at KMTC from the beginning and knew all the Afghan officers and NCOs who made up the KMTC cadre. He was very close to General A, the Afghan commander of KMTC and he was the TAG commander's interpreter. Because General A spoke English, JS was able to spend most of his time supervising the other interpreters and the translation work.

The interpreters, known as 'Terps' by the slang producing SF troops who had originally hired these men, were critical for the success of our mission. Colonel Davis said later that they were the 'center of gravity,' a doctrinal term made popular by General Schwarzkopf during the Gulf War, for our effort at KMTC. It was apparent to me from the beginning that the personal relationship between the American soldier and his interpreter was an essential element in the development of effective relationships between us and our Afghan counterparts. It was as important for our guys to establish rapport with their interpreters as it was for them to do the same with their Afghan counterparts.

Some of the comments I was hearing from the Vermont guys led me to believe that the people at Camp Phoenix who held the purse strings felt that interpreters were plug and play items who could be used by a different person every time a need for inter-action between us and our ANA counterparts came up. The Brigade staff had very little interface with the Afghans, so it was easy to see how they could develop this kind of attitude. It made sense to my cynical outlook about 'Intel' that the S2 would be the overseer of interpreter operations because the S2 was about as ignorant as they come of the fine details necessary for working in this environment. I was pretty sure that the deputy S2, Major M would have been much more sensitive to these things but he was not in the driver's seat and as I came to understand, not his father's favorite son.

I was determined to have an interpreter for each of our people who had to have a direct contact with an ANA officer or sergeant. I wanted each interpreter to remain with the section he had been working in with the Vermont guys because he already knew the ANA people and he would be able to help our guys to establish rapport. Most of these interpreters were pretty smart and they would take care of a lot of the cultural pleasantries that we wouldn't even think about. We had been trained to put the bottom line up front and the Afghan culture was just the opposite. All of these things needed to be worked out between the American and his interpreter before effective communication could begin with the Afghan Army people we were there to help.

Bill and I and the Interpreters-KMTC
(author photo)

Jalalabad Road Driving Qualification—OJT

After a few days, the Vermont LTC told me I was driving. He rode shotgun and directed me to the front gate where 'Rambo' gave us a nice salute. I pulled up to the edge of Jalalabad Road and looked for a break in traffic. The LTC told me to just drive into the mess and people would get out of our way. I wasn't sure if he knew what he was talking about or if he was just suicidal so I crossed the first lanes of traffic to our left and waited in the center median for another break which came soon enough.

I floored the SUV and now we were driving down Jalalabad Road as fast as traffic would allow. There had been a few bombings on this road; back in June four German soldiers had been killed when a suicide bomber rammed their bus nearby. I didn't see how we could avoid a determined attack like that so I gave myself to the Lord and did the best I could to avoid all the obvious hazards and let Him and the others watch for possible attackers.

The traffic up ahead appeared to be slowing down and then it suddenly stopped. I didn't want to be stopped and surrounded by a lot of people I didn't know so I swerved to the left into the dirt median and pressed on through the dust cloud left by people ahead of me who had also felt the need to press on. The LTC sitting next to me had a white knuckle grip on the overhead hand grip and was looking at me with a kind of, 'do you know what you're doing?' look, but he didn't say anything as I strained to see what was ahead in the dust as we bounced up and down passing the stopped traffic on our right.

The center divider strip was generally smooth dirt but it usually had people standing on it, vehicles parked on it, and there was the occasional bomb crater. After we passed the obstruction, a truck stalled sideways across both lanes, there was a clear highway. I swerved right and got back on the road and we continued on to KMTC. The LTC complimented me on my driving and after that I considered myself 'Jalalabad Road Qualified' but I had a long way to go.

Driving on Jalalabad Road would be the most dangerous thing we did in Afghanistan. Several days later four people from Rhode Island rolled their vehicle. They were run off the road by a bus that swerved to avoid a taxi that was pulling onto the road. They hit one of those bomb craters, rolling their SUV. One of them, the driver, received a laceration on his scalp from the rifle carried by an LTC who was sitting in the front passenger seat. The husband and wife in the back seat were badly shaken and the wife split open her elbow on her window's track. She and the driver were taken to the ISAF hospital at Camp Warehouse by British Army personnel who were passing by at the time of the wreck.

They were accompanied by the LTC who was riding in the front seat and a first sergeant who was riding in the vehicle that was trailing the wrecked vehicle. Several others provided security and assisted the sergeant who had been in the back seat as he recovered the driver's rifle which was pinned under the SUV, now resting on its roof. These were the first injuries suffered by our unit in the great war on terror but they wouldn't be the last. Someone involved in the wreck later attempted to have purple hearts awarded to the ones who had been injured claiming that the taxi driver was a terrorist.[79]

The accident would also highlight another problem that would plague us through the entire deployment, no vehicle radios. There was no way to contact Camp Phoenix for help. A passing U.S. unit was able to radio the Ops center at Phoenix and they sent the QRF (quick reaction force).[80] I was used to instant communications in peacetime CD operations and here we were on the other side of the world in the largest deployment of U.S. forces since the Gulf War and we couldn't talk to our base that was less than five miles away. This was only part of our communications problem and we were forced to use our own commercial walkie-talkies for short range communication around KMTC.

The White Ford Ranger

After a few days we were told to turn in our SUVs and we were issued White Ford Ranger (WFR) pickup trucks. They had a four door crew cab and diesel engines. Before long, every American military unit in Afghanistan was driving them. While I was there no WFR was attacked by the enemy. They were easily the most recognizable vehicle on the road with no tactical features. There was an effort to put a plywood wedge in the back of each one to prevent hand grenades from exploding in the bed causing them to roll off instead; but not all the vehicles were ever so equipped.

There were quite a few wrecked WFRs in the KBR maintenance yard but they were all wrecked by their drivers. Afghanistan is the land of conspiracy theorists and I considered the WFR as a multi-level deal: the company that sold them, the Taliban who didn't want to stop American aid that was pouring into the country by accidentally killing an American soldier, and the Afghan government which was pulling some big strings behind the scene.

Guns and Poses

I began to settle into my office which was downstairs and away from the rest of the staff upstairs. I shared the office with the medics:

a wild eyed Hispanic captain from the Medical Service Corps, an older Master Sergeant, and a young Specialist medic, all from Texas. There was another Master Sergeant from Rhode Island named Bob who was a Rhode Island State Trooper in his civilian job and there was an office next to my desk which was occupied by another member of the staff, Major M.

I was still carrying an M-16 rifle around and it was a drag. I had to spend a lot of time with the interpreters and I felt like the rifle was an intimidating presence that I didn't want. It was also a hazard because I had to take it off my shoulder where I kept it slung when I was walking and had to hold onto it or lean it against the wall whenever I sat down. If the interpreters decided to take me out they would have an automatic rifle to work with.

I went to Major Shawn, the G3 Ops officer and told him I needed a pistol and a few days later he told me to turn in my rifle and draw a pistol from supply. When I went back to the tent that night I heard the medical specialist complaining that his pistol had been taken away and now all he had was an M4 carbine; he wasn't happy when he saw me wearing a pistol on my waist later that evening. I chose to wear a pistol belt and an issue holster that I had brought with me rather than an SAS style 'drop rig' thigh holster or a 'FOB bra' shoulder holster worn by everyone who didn't have an SAS holster. The drop rig thigh holsters have been useful for storming an embassy but the pistol was almost impossible to get at if you were a passenger in a vehicle.

I felt a lot more comfortable with that pistol but many of the rest of the troops wanted more fire power so they managed to get their hands on some AK-47s that were floating around at the Alamo. There was a practical reason for this. At Fort Carson we had been issued a few M4 collapsing stock carbines, many M16A2 full sized rifles, and a few M9 pistols. They had been distributed without any rhyme or reason I could discern. Of course most of the officers ended up with the pistols and M4s. The full sized rifles were more than useless for shooting from the inside of the cab of our small pickups.[81]

Soon all the security advisors and many others were carrying folding stock AKs which were much more practical for firing from inside a vehicle which was the most likely shooting scenario that I could envision. Everyone who carried one went to the range one day and qualified. Of course they felt the AKs made them look cool in a camp where most troops never went outside the gate.

This went on until some of them were spotted by a General up at Bagram Air Base. This happened shortly after Ranger Pat Tillman was killed by his own men in a 'friendly fire' incident. The General, who worked at OMC-Alpha, considered that there was the potential for a similar incident and contacted the 45th Brigade General and ordered him to get rid of the AKs.[82] The guys had to stop carrying them but most of them kept a few under the seats of their vehicles until they went home. Of course, the 45th Brigade never offered to help out with the weapons or the ammo shortages.

Hearts and Minds

Since it was winter and I'm not happy when I'm cold, I began to observe the interpreters and how they dressed. There was snow on the ground for a few days each time it snowed and puddles of slush everywhere. The first thing I noticed was that most of them were wearing sandals. I assumed that they were poor and that was all they could afford. Then I noticed that their jackets were thin and didn't seem sufficient for the cold temperatures we were experiencing. I wondered how they could function in the cold weather.

My wife emailed me with some information from a sister in the church who knew some women who wanted to send things to the troops. They were making and sending some kind of sweat cloth that you wore around your neck to keep you cool, not exactly what we needed in Kabul that winter. I wrote to their leader and asked if she would be interested in helping to clothe my interpreters with warm jackets and cold weather boots. She replied that she and her friends would love to help and asked me to send sizes. I ran this by JS and he said that I was nice to think of such things. I asked him to get me shoe sizes and he said he would. I continued to correspond with the woman and found out that an executive with a parcel delivery company had offered to ship the boots and jackets to their office in Kabul for free. Now I began to think that we might be able to do something good for these interpreters who were crucial to our mission.

Military Professional Resources Inc.

Outside our office in the hallway were some offices occupied by MPRI, a private military company that had a contract to write doctrine and training products for the Afghan National Army. One office had several retired sergeants who never spoke to us and I never got to know any of them. The other office was occupied by a retired LTC and he gave me a lot of information about Afghanistan. I asked him if he would give a briefing to our people and he did.

He gave us information about personalities in the ANA. Then he covered certain aspects of Afghan culture and manners which would be useful for our people in their attempt to establish rapport with their counterparts. He told us about Chai, the Afghan tea and the role it played in meetings; basically you should be ready to drink tea every time you visited an Afghan. Finally he dropped a bomb on us when he said that farting was one of the worse things you could do in the presence of an Afghan.

There were some doubtful expressions on the faces of some so he gave an historical example of a U.S. diplomat who noisily bled off some gas at a meeting and none of the Afghans would have anything to do with him afterwards. This information would prove to be very correct. However, as the U.S. has been there for ten years now the Afghans have had to modify their response to our bad manners but they still hold the ones who display such bad manners with contempt in their hearts. A few had been known to kill for what seemed to be trivial offenses and with the insider attacks increasing, I wonder.

The MPRI man had just missed Vietnam but he had served with the only mechanized airborne infantry unit in the U.S. Army, the 509th, part of the 8th Infantry Division in Germany. I had met some soldiers from that unit when I was a TDY guard at the Mannheim Stockade in 1966. When I told him he was surprised because he had been accused of lying anytime he told people about that unit and he was pleased that I knew about them. We had some good discussions about the army of our youth. When I told him about my training experience of climbing a troop ladder into the back of a Chinook and then climbing down on another hanging from the center of the aircraft, he called bullshit, so I found a picture of one on the internet and showed it to him and after that we got along even better.

The Twelve Interpreters

One of the first things I had to do was sort out the interpreter situation. Each member of our unit needed an interpreter in order to talk with his Afghan counterpart. All the remaining Vermont troops handed off their interpreters to their replacements. Since many of the Vermont contingent had already left, I had to get with MSG Bill and JS in order to arrange meetings with our guys and the interpreter assigned to them.

After figuring out how many interpreters we had and how many we needed, I created a spreadsheet for my Power Point slides so I could brief the command group. In the process of doing this I discovered that we had twelve interpreters who were working for us without pay.

MSG Bill had been struggling with the 10th Mountain Division and now with the 45th Brigade about the need for more interpreters for KMTC. He had decided on his own initiative to hire the next twelve 'walk-ins' on a 'trial' basis. He had done that earlier and was able to add ten interpreters who were eventually paid out of his unit's funds until they were officially hired by TF Phoenix. This time he had not been able to get anywhere and he kept telling the twelve Afghans that they would be paid 'soon.'

I wrote a memo to the CO about the need to hire more interpreters, starting with the twelve and the need for back pay. Then I went to Major Bobby, the Brigade S2, thinking that this was a reasonable request, and told him that I needed to hire these twelve and get them their back pay. He jumped out of his chair and told me that not only were they not going to hire the twelve and give them back pay, but to, "Fire them today!"

Obviously that didn't sit well with me. Here I had been preaching rapport building and all the other special ops rules, which I had been practicing for the past thirteen years. Now in Afghanistan, a place that screamed for the application of those rules, a field grade officer was telling me to fire twelve men, all educated in the University, many of them doctors and engineers. They were all young Afghan men who had been working without pay on the promise of a U.S. soldier. I couldn't think of a better way to add recruits to the anti-coalition insurgency that was simmering there in the winter of 2004. Realizing that arguing with Major Bobby was a lost cause, I left without saying anything and called on the Lord under my breath as I walked back to my tent.

SSG Carl, one of the artillery guys, became our IT guy at the Alamo. The SF had left a satellite connection to the internet and SSG Carl figured out how to use it. He ran a cable down to my desk and I was able to get on-line. One of the first things I did was to contact my wife by email and ask her to pray for the situation with the twelve interpreters. I ended up sending her pictures of the twelve and she distributed them to her prayer partners in Austin and they began to pray.

Early Morning Routine

Every morning at 0330 I would wake up and go to the Brigade intel office for my morning update. I would walk through the camp and thought it would be very peaceful if it wasn't for the sound of 50 diesel generators roaring all around me. I was starting to get to know the night shift people and they were very helpful.

At 0500 I could barely hear the local mosque mullah calling for morning prayers over his PA system. I was usually done with my intel update so I would head for the DFAC and another KBR breakfast. Usually Matt the friendly young local was there at the DFAC entrance, helping people with signing in and keeping the area clean. The first time I noticed him he asked me the meaning of an English word he had written on a piece of paper. I tried to explain the meaning and found out that he was teaching himself English. He had a friendly smile and so I looked forward to seeing him each morning.

The Dogs of War

Going to work was fun in a weird, NCO way. Sergeants have to be problem solvers and there was a conveyor belt of them at KMTC. One of the next problems I had to deal with was the dogs. The SF guys had taken in a few local dogs and trained them to bark whenever an Afghan came around them. How they did it I have no idea but these dogs would never bark at an American in uniform but they would bark and snarl at the Afghans that came to the Alamo. The MP's were amused by this and the dogs took this for approval and began to snap at the interpreters as they walked through the courtyard.

The dogs were quickly adopted by some of the dog loving Rhode Islanders and so I tried to reason with them regarding the dogs threatening to bite the interpreters. They thought it was funny so I told the interpreters to stay together when they walked across the courtyard. One morning I heard a big ruckus outside. There in the courtyard was a young Afghan boy who had been hired to do janitor work at the Alamo. The dogs had him surrounded and one of them was lunging at him and trying to bite him. The kid was terrified. I saw a couple of the Rhode Islanders watching all this with amusement but I was so mad at this spectacle that I abandoned any attempt at rapport building with my East Coast comrades. I walked quickly and quietly across the courtyard and got behind the aggressive dog where I wound up and kicked him in the butt as hard as I could.

As he flew through the air I drew my pistol just in case he took a dislike to me. I didn't have to worry though because at the end of his trajectory his head struck a pillar and he let out a second yelp before crumpling to the ground. One of the Rhode Island sergeants came over to me with a red face and began to berate me for kicking a defenseless dog. I went off on him with, "what's more important to you, dogs or humans?" He didn't say anything so I went on, "The next time I see these dogs attack one of these Afghans I'm going to shoot all of them!" I was actually looking forward to making good on that threat when an order came down from General Mancino at 45th Brigade directing that all animals be removed from any Task Force Phoenix facility. That was one of the few directives from them that I appreciated.

Wish in One Hand...

The IT guy, SSG Carl told us to put in an equipment wish list so I spent a few hours looking up computers, cameras, and software, thinking that I would create some training videos for the ANA in my spare time. I turned the list in and waited for my stuff to arrive. I guess that list joined the one I had made previously for mission essential equipment to be used by us on our night combat patrols that Colonel Knight had told us about so long ago. That list and my computer list bore no fruit. Fortunately, the CO saw that the night combat patrols were unnecessary and put a stop to any more attempts by Colonel Knight to get us into a fight with someone in the dark.

The Afghan Dilemma

A few days after the dogs got kicked out, I was sent to the front gate to take a report from the MPs regarding an Afghan civilian who had been beaten and arrested by an ANA sergeant. The sergeant had turned the civilian over to the MP's with a report that the civilian was trying to get ANA uniforms and weapons which he would use to kill all the Americans at KMTC.

The ANA sergeant was gone when I got there so I took the civilian from the MPs and asked him, using the MPs interpreter, what he was doing there. He told me that he was a contract employee at KMTC. When I asked him why the sergeant had beaten and arrested him, he said that he had said good morning to the sergeant in English and the sergeant began beating him and told him that he had to speak Dari at KMTC. The civilian's employer arrived about that time and vouched for his employee so I took his picture and let him go. I saw several instances of this 'language training' while I was there, usually one of the Tajiks insisting that a Pashtun speak Dari rather than Pashto.

One of the Vermont SOPs had required MSG Bill to escort any interpreter coming onto the compound. Every ten minutes I would get a call on my radio from the MPs at the gate advising me of an interpreter needing an escort. I went to LTC Randy Davis with a memo changing this policy and from then on, after the interpreters were searched they walked by themselves to the interpreter office where they would be picked up by their American counterpart—one more hassle out of the way.

Back at Camp Phoenix one day, I was reading the camp newsletter. I decided to submit a small column on the New Testament revelation of God's Economy. I made my submission but my proposed column was rejected because the chaplain had that territory covered. I resolved to find the chaplain and attend his Bible study when I could. That would prove to be a mistake.

The Brigade S2 Probes

I got a demand from Major Bobby. He wanted to know what exactly we were translating at KMTC that required so many interpreters. I had already had several negative contacts with this officer and he was working hard to squash me. I mentioned this to my wife and requested that she and her prayer partners continued to pray for the interpreter situation. I wrote the Major a memo detailing the translation work that had been done and what was in-progress and I didn't hear any more about translation for a while. Meanwhile I continued to bombard the CO with Power Point and memos detailing the need for interpreters, the twelve unpaid interpreters and their back pay.

Rapport Building Road Blocks

I had emphasized rapport with the interpreters several times in classes back at Fort Carson and most of the troops worked hard to establish it with their interpreters either due to my exhortations or their own natural good nature. There were a few that just didn't get it and they were mainly from Rhode Island. Many of the Texans were 'good ole boys' and some of the Rhode Islanders talked like 'wise guys' from some movie about the mafia.

One of the RI sergeants came to me and complained that he couldn't get his interpreter to curse at the ANA trainees. I told him that cursing the Afghans was a bad idea but he was having none of that. He was a big guy and worked as a guard in a federal prison back home. I'm sure the prisoners were glad to hear that he had deployed to Afghanistan and were probably rooting for the Taliban.

When I asked JS about the issue, he said that the interpreters didn't want to anger the Afghan soldiers because they were afraid of getting beat up or worse if they ran into them outside the camp. He told me that Afghan culture puts personal honor above almost every other consideration and insulting an Afghan in front of others was a sure way to get hurt. The Rhode Island sergeant kept pushing his interpreter to teach him how to curse in Dari until the kid quit.

I had to find an older man who was able to convince the sergeant that he was translating everything that was said while toning down the bad language to a level that the Afghan soldiers could tolerate. Actually he told everyone that the sergeant was a bad mannered person and that because the Americans were helping them they would have to ignore the bad manners as a small price to pay for our help.

Pay for the ANA

The pay for the ANA was handled by us. We provided the money and supervised the ANA officers who actually paid the soldiers. Payday was a like the paydays of my youth in the Army. Long lines of soldiers at the headquarters waited impatiently to get their money and when they did they counted it several times. The soldiers had been paid in U.S. dollars and now the ANA officers had heard that the pay would be in Afghanis, the Afghan currency. The official rate of exchange was 50 Afghanis to the dollar but the Afghan confidence in their country's currency was low. I alerted the staff at the morning briefing and advised them to keep their ears open to their counterparts. When the next pay day came we were a little more alert but there were no incidents and it quickly became a non-issue.

Terrorists At KMTC

The next incident would take up a lot of my time and is a good illustration of the difficulties faced by our troops as they try to build an army made up of young soldiers who were coming out of the chaos of that country's history. I got a radio call to come to the detention cell at the Alamo. There I was met by SSG R, one of the RI soldiers assigned to basic training. He had observed one of the ANA recruits making a sketch of KMTC during a break. The ANA trainees were on a hill in the middle of KMTC and this soldier was sitting in a group of soldiers happily drawing his sketch when SSG R grabbed him and placed him in detention.

The ANA officers from his Kandak searched his bunk in the barracks and found several pieces of paper with Arabic letters. I had one of the interpreters translate the documents while I interviewed the trainee. The trainee told me that his group had just had a class on sketching and that he was practicing what he had been taught. He seemed to be amused that he had been arrested and seemed sure that we would see that he was innocent and return him to his comrades.

The interpreter gave me a translation of the first document. He said it was a poem attributed to Osama bin Laden. The poem declared that the writer had dealt with the Najibullah government and now was on his way to Moscow to deal with them, chanting Allah Akbar. This didn't look good. The next document was the cover of a notebook. It had a dedication declaring it was a gift from Ayman al Zawahiri, then thought to be OBL's right hand man. Things were getting worse. I told the young recruit that we had found certain documents that made it appear that he was a member of Al Qaeda and he laughed and said that there was a joker in his unit that had written those things for a laugh. All my police experience led me to believe he was telling the truth.

The RI SSG was sure that he had captured a major terrorist and hoped to get a medal for it. The ANA officers were ready to lynch the kid so I recommended that he be detained and sent to Bagram as a 'PUC' (person under U.S. control) and let the interrogators there figure it out. I told him that it wasn't a joke now and that he should tell the truth when he was questioned. He was still friendly and seemed to think that this would be cleared up soon so he could get back to basic training. That was wishful thinking on his part.

Stripes

Since the TAG had been thrown together with soldiers from several states and many different units, the positions on the manning roster were filled with personnel with as close to the appropriate rank as possible. In my case it was a stretch because the S2 would normally be a captain and I was a master sergeant. There were three sergeant major positions and each one of these was also filled by a master sergeant, the actual sergeants major from our units back home were unable to deploy for one reason or another.

The man chosen to be the command sergeant major apparently felt that he wasn't getting the respect he felt his position deserved and the CO directed him to wear CSM stripes instead of the MSG stripes he was authorized by regulation to wear. This was called 'frocking.' I had seen it done here and there over the years and didn't pay any more attention to the 'CSM' with his new stripes than I did with his old stripes. Our duties had no overlap and I seldom saw him. The other two sergeants were told to wear sergeant major stripes and they refused, citing the regulation that prohibited this action. The CSM called a meeting with them; he told them they were 'frocked' and to get with the program and wear the stripes. They refused. This made the CSM even more unhappy and he must have made his feelings known to the CO.

Then LTC Randy Davis asked the sergeant working for him to wear the stripes and he complied. The other sergeant refused the request of his LTC and asked to talk to the CO but was denied an appointment. He continued to refuse to be 'frocked' and was advised to carry a copy of the regs wherever he went and he did. All this seemed to increase the tension in the senior NCO ranks but I was blissfully unaware of most of it.[83]

More Terror at KMTC

Two days later the CO told me to contact one of the Corps of Engineers (COE) people that were working on the water system at KMTC. The engineer had been wandering around in some bombed out warehouses in an unused part of KMTC. There he found a large room full of Russian heavy machine guns in good condition watched over by some men in older Afghan military uniforms. This caused a big hub bub and our intrepid CSM decided to do a personal reconnaissance to gather more information about this threat to KMTC.

When he came back I objected to his endangering himself and the rest of us by stirring up trouble with the Northern Alliance troops that were staying in that area. The CO then sent me to the same place to verify what the COE people had reported. I went with the security mentoring guys and looked through an open door at a long row of wheeled anti-aircraft machine guns that appeared to be in good condition. I heard some footsteps so I walked back to the vehicle and watched as a bearded man in uniform carrying an AK came to the door and looked around. We made eye contact and I waved and gave him my friendliest smile. He smiled and waved before closing the door.

The area around KMTC had been a Soviet Armored Brigade headquarters during the jihad. When the Russians left Afghanistan in '89, they left behind much of their weapons and equipment for the Najibullah government. These weapons had been used by the Russian puppet government until it fell and then by the mujahideen in the civil war that continued until the Taliban made their appearance. Since many of the Northern Alliance were members of the communist government's army, they kept these weapons and used them in their fight with the other mujahideen groups and then with the Taliban.

Even though the Transitional Government was now in charge, the Northern Alliance people were holding on to their guns in case things went sour and this was what the COE guy found when he was snooping around where he shouldn't have been. There was a disarming program going on but that wasn't part of our job. I recommended that we ID the location and pass the information to 45th Brigade, while at the same time telling our troops and anyone else working at KMTC to stay away from that area. The CO agreed and we heard no more about it.

The Interpreters

Each morning I had to prepare my slides for the staff meeting. I kept my part of the briefing as short as possible while making sure that the interpreter situation stayed on the front burner. After the meeting I went to the interpreter office and spent time with JS. Every time I walked into the crowded room, all the interpreters would stop what they were doing and pay homage to me. This, I felt, was ridiculous. I tried to tell them that while I appreciated their friendship, I thought it was a waste of time and energy to stop everything for 15 minutes while thirty interpreters shook my hand and hugged me. Try as I might, I never could break that custom so every day I did the 'greet and grin' with them. I did enjoy being with them but as soon as we were done with the formalities I would sit with JS and go over the list of problems for the day.

There were usually several interpreters who needed a few days off to take their pay home. Most Afghan men live in their father's house. Even after they marry many of them continue to live in their father's house and they give their pay to their father who distributes the money collectively. I would usually let JS decide if they needed to leave and as long as another interpreter was available and the American counterpart was ok with it I let them go and covered for them on the payroll. JS would remind me that Eid was coming and that he hoped we could get the pay for the twelve before Eid because it was like an American Christmas with everyone buying presents and getting new clothes to wear for the family celebrations. Since I didn't celebrate Christmas I was a little less sensitive to this issue than I should have been but I did press the issue every chance I got.

JS and I came up with a current figure for the unpaid interpreters, $3900.00. This was a bit more than I was prepared to pay to keep our honor clean but I was determined to do everything necessary to make the money boys at the Office of Military Cooperation-Afghanistan (OMC-Alpha) pony up.

JS and I would sit around talking after we had gone over the pressing business. He and some of the others would ask me questions about where I was from, was I married, did I have any children, what I thought about Afghanistan, etc. They were tap dancing around what they really wanted to know which was what I believed. "Are you a Christian?" one of them finally asked. I told them that I was and this began a number of question and answer sessions that would go on until I left.

45th Brigade Counterintelligence

The CI team was feeling special because as counterintelligence agents, they were allowed to wear civilian clothes and they had relaxed grooming standards. They were both growing beards and they couldn't have been more incongruous as they accompanied the 45th Brigade local presence patrols in their 511 fashion show clothing and sunglasses. They were supposed to develop local agents who would keep them informed about possible threats to Camp Phoenix but they had not been given any Op funds for paying informers. The Afghan spies are a market oriented people and information products were always being offered but they couldn't do any buying because of some bureaucratic red tape at Bagram Air Base.

The CI guys let me know that they were the only ones with Task Force Phoenix authorized to do any collecting. I asked them to stay in touch because I would give them any information that came my way. They told me that KMTC was part of their jurisdiction but things at Camp Phoenix kept them busy and reminded me again that they were the only ones authorized to collect. I assured them that I understood but mainly I understood that these were young soldiers thrust into an environment that was a lot different than the school house they had attended. Their arrogance was what put me off and in the months to come I would be frustrated several times in my attempts to work with them.

The TAG Newspaper

One morning at the staff meeting, the CO dropped a bomb on my position when he told the staff that the S2 (me), would now be writing a daily newspaper for the unit. He said that he was concerned about the troops losing touch with things back home. This really angered me. As if I didn't have enough to do, supervising some 30 Afghan interpreters and trying to unscrew the mess left by the last unit, who promised money for work and then left town. If they had done it for personal gain I would have been hammering on the door of CID, the FBI, or whoever was handling criminal corruption cases in Afghanistan but they did what they had to for the accomplishment of the mission and now they were gone.

I went to the CO after the staff meeting and told him that we had a PAO who wasn't especially busy. He didn't look like that was information he wanted to consider. The next day at the staff meeting when he asked about the newsletter I told him that I had checked with a good sampling of our unit and no one expressed any need for more news from the home front. Then I attempted to slam dunk the whole argument by saying, "Besides, who gives a shit about how the Patriots do in the Super Bowl."

That was a dumb thing to say to the Rhode Island crew. LTC Kevin let me know how he felt about this critical event that was apparently taking up a lot of his considerations, by jumping up and storming out of the room. That was unusual behavior for an LTC in a staff meeting but this was war. I was beginning to see how strategies were developed and decisions were made on a staff—those things that caused both laughter and cries of anguish from the troops, depending on your view of life in the military.

The CO directed me to go and apologize to him after the staff meeting. I prayed for strength and went to LTC Kevin's office. He was sitting at his desk with his head in his hands so I asked him if I could speak to him outside. He got up and followed me outside where I pleaded with him to forgive me for offending him. He said he would and we went about our day with me thanking the Lord and wondering what was next.

The English Classes

I didn't have long to wait. JS told me at our morning meeting that four interpreters had been teaching classes to the ANA officers and hadn't been paid. This was another program set up by the guys from Vermont and now they were gone. JS told me that General A, the ANA commander of KMTC was very pleased with these classes and so I went with him to talk to the General.

General A was an older man who had been in the Afghan Army before the overthrow of the King in 1973. He had studied in the U.S. for several years attending U.S. Army schools. Because he had been to the U.S. he wasn't trusted by the communist government when they came to power in 1978 so he remained a captain through the war with the Soviet Union, the civil war, and the Taliban war that followed. When the U.S. came to Afghanistan he was promoted to General and became the father of the modern Afghan National Army.

JS knew the General quite well and got us in to see him. General A told me how he liked the English classes and wanted them to continue. This gave me the ammunition that I needed to write yet another memo to the CO requesting that we get funds to pay our debts and continue our programs. The General was a very pleasant man and I was to have a number of dealings with him over the next 6 months. He saved me from a lot of trouble when my ill-advised KMTC Intel Estimate upset some of his officers.[84]

The 45th Brigade 'Good Idea Fairies'

The staff people at 45th Brigade were bored. The 45th had one battalion pulling security at Camp Phoenix. Its other battalion was broken up and the senior officers and NCO's were formed into Embedded Trainer Teams to act as advisors to the ANA. These teams were OPCON to the SF and they were far from Kabul.

The staff felt left out of the usual opportunities for brilliant staff work because there was nothing really going on. They decided to get some attention from the General by having all the people who were actually working do more work. This meant that all the functioning elements of TF Phoenix had to come up with briefings to the General on how we were going to improve the effort we were involved in.

I was called to a meeting with the Central Corps Advisory Group, part of Task Force Phoenix, where I was assigned to brief the General about interpreters and what we could do to establish/improve the whole interpreter program.

We would have weekly meetings led by a Major M, a reservist who managed to get on this deployment. He wore a Ranger Battalion combat patch and very much wanted to 'lead' this project to the point of telling us what he wanted us to say. Of course this didn't go over well with me and I continually resisted his attempts to screw with my Power Point slides as we went along. We were all being polite to one another but as I listened to what was going on at Pul-e-Charkhi, where the advanced individual or MOS training was taking place I started to question what exactly we were trying to do in Afghanistan.

What Are We Doing Here?

The advisors there were taking U.S. Army doctrine; techniques, tactics, and procedures, and using it to train the Afghans. Apparently no discussion had taken place with the Afghans and if there was it was a discussion by people on crack. One of the colonels at KMTC gave me the perfect illustration for this when he complained that we were teaching the Afghans to zero their AK-47s at 25 meters but they had always zeroed them at 100 meters, just like the Russians taught them.

I'm pretty sure that the higher ups, when they complained about not having any input and were told which side of the bread the butter was on, decided to go along with the Americans for as long as we were here and then they would deal with what came next. We had been sent to Afghanistan to raise up an army and because we were in a hurry, we told the Afghans who were depending on us for everything, that they needed to train their army to be just like ours. I remembered us doing the same thing in Vietnam and became like a voice crying in the wilderness.

When I questioned the wisdom of teaching them things that they could neither understand nor sustain after we left I was told to shut up by the ones who outranked me. In a delusion of grandeur I wrote a letter to the major specifying my concerns and offering to school the leadership on certain principles of unconventional warfare and foreign internal defense but received no rely.[85] I did notice the major looking at me a little more intently; maybe he was told to keep an eye on me.

Pul-e-Charkhi

There was another coalition base to the east of us called Camp Black Horse. It was in the middle of the large ANA base at Pul-e-Charkhi. The ANA Central Corps headquarters was there along with a number of advanced training schools for the ANA infantry and armor forces. I had to go there occasionally when someone needed another body for a two vehicle-two long gun-four people convoy. There were two ways to get there: through the ranges or Jalalabad Road.

The range road took you through the Bone-Yard. There had been a 'battle' here just before we came and it was a Russian vehicle diorama. Hundreds of T-34s, T-54/55s, PT-76s and T-62 tanks, BMPs, BRDMs, BTRs and other machines from the Soviet war machine littered the landscape. Just passed it was a road to 'Bin Laden's cave', a large tunnel cut into the side of a hill. It connected with a cave inside and went all the way through to the other side. One side had been bombed but you could drive in the other side. Maybe this was the type of cave Rumsfeld was talking about during the battle at Tora Bora.

When we took Jalalabad Road it was a lot faster but you had keep off the shoulders because of the mines. There were some non-government organizations (NGOs) doing mine clearing work here and you could keep up with their progress because they painted the red rocks, which denoted a mine, white, when they cleared it. About a mile south of Camp Black Horse was the infamous Pul-e-Charkhi Prison, once the largest prison in Asia.

They say it was a slaughterhouse during the communist years and now it was being refurbished for another stint as a maximum security prison by the new Afghan government. Soon it would be holding 'Jack' Idema, a controversial guy who figured heavily in Robin Moore's book, "The Hunt For Bin Laden."[86] He was in Afghanistan looking to claim the $25,000,000 reward and was arrested at a house in Kabul where he and his buddies were torturing some Afghans for information leading to the pot of gold.

Bill, the man I replaced at KMTC was at Camp Black Horse and I would track him down to say hello from time to time. He always had some new project that he wanted me to buy into so my visits were short. He showed me around the camp though and I was struck by two things; the enormous number of U.S. Army five ton trucks that probably come from every National Guard and Army Reserve motor pool in the states and the Scud missiles that were parked near the tanks and BMPs of the Afghan armored units. The Scuds had come from the Panjshir Valley recently and I wondered how many more there were out there.

Kabul Military Training Center
(author photo)

Clothing Issue to Recruits
(DoD photo)

Shoe Sizing
(DoD photo)

RPG Training
(author photo)

PKM-Russian Machine Gun and AK-47s
(author photo)

Kandak Graduation
(author photo)

But to Him who is able to do superabundantly above all that we ask or think...
Ephesians 3:20

Chapter Four: February 2004

The Camp Phoenix Bazaar

Friday was our day off because it is the Muslim day of prayer. It is also a day for business and we had a bazaar at Camp Phoenix every Friday. The 45th Brigade troops at Camp Phoenix were so overwhelmed by their duties that they made us pull security at the bazaar once a month. Since security has something to do with intel according to the thinking of those who haven't read the manual, I was given the task of Sergeant of the Guard and directed to come up with a duty roster.

The CO came up with a nice touch on this one and directed that the senior officers and NCOs would pull this duty, maybe remembering Colonel Knight's fantasy that we would all be pulling combat patrol duty. I dutifully made up a roster for hour long shifts with the uniform and equipment, posted it in several places, and briefed it in the morning staff meeting several times.[87]

When I came to the bazaar early on Friday morning to make sure that we were doing our part, I was fascinated to see the goods of central Asia being unloaded and moved to the small space allotted to each seller. I wandered through and looked at what was for sale. There were a lot of old British rifles, some were real and others were made in a village in Pakistan that specializes in weapons manufacture.

There were large stacks of carpets from all over. They were quite expensive and the people selling them were a study. They had stories about the origins of the carpet that were fascinating even if they were not true, but who knew? You would usually have to get one of the Camp Phoenix interpreters to help you with the bargaining and I knew that some of them were connected with the vendors and got a cut of whatever sale they were able facilitate.

I got to know one of the interpreters who was a doctor, working for more money than he could make practicing medicine. He was famous for saving the lives of several SF sergeants in the early days when they were attacked in downtown Kabul by someone who threw a hand grenade into their vehicle. The interpreter treated their wounds and got them to safety even though he too had been wounded. I kept waiting for Camp Phoenix to give him an award and do something for him but it wasn't until much later that he qualified for a special immigration visa and came to the U.S.

The most popular vendors were the DVD sellers. They were selling pirated copies of the latest movies and computer software, all made in Pakistan. Some of the movies had been shot by someone with a video camera in a movie theater. It startled me the first time I saw one of these movies because although the sound wasn't very good I thought I was watching a poorly made DVD until someone in the theater audience got up and walked across the theater in front of the camera.

At the end of the day I was standing with the last guard shift watching as the shoppers made their last purchases and filed through the checkpoint. A group of people in civilian clothes approached and I could see that they were wearing U.S. Embassy ID cards which were hanging around their necks. One of the men in the group stepped up to the sergeant on guard and said, "I want to talk to whoever is in charge here."

The sergeant nodded towards me. "Yes sir," I said. "How can I help you?" "Do you know that they are selling a tiger skin in there," he said, waving his hand towards the bazaar. "Tigers are an endangered species and I'm going to report this," he said with all the assurance of a man who was sure that everyone saw things the way that he did. He stood there looking at me like I was supposed to do something and when I didn't, he demanded "aren't you going to do something?"

Here was this diplomat with a large bag of pirated DVDs and software demanding that I do something about an endangered animal, much too late for the one who's skin he saw somewhere in the bazaar. "Sir, we're in a country where the humans are endangered and we're doing what we can to stop that killing and until we do, as far as I'm concerned, the tigers are going to have to take their chances like everyone else." His face got all red and he looked around for some fire support but the people he was with had moved on and he wasn't finding much sympathy from the soldiers with me so he scurried away with his goods.

Eid al Adha

The matter of the twelve unpaid interpreters was coming to a head. As Eid al Adha, the Muslim feast of sacrifice approached, JS told me that the unpaid interpreters were getting anxious. This holiday meant a lot in that culture and they needed money to take care of their families. I still wasn't getting any joy through official channels and I stayed away from Major Bobby at 45th Brigade after he told me to fire the twelve three weeks earlier. I had been on the phone and email with my wife and eventually the whole church in Austin prayed for my burden to hire these men and get them their back pay.

I went to finance and got two hundred dollars. I had some other cash and I gave each of the twelve a $20.00 dollar bill. I thought that would be big money in Kabul but instead of getting a big response of gratitude, most of the twelve took the twenty dollar bill and didn't say anything. One of the younger ones took the bill, looked at it and then laughed in my face before storming off.

Another problem arose. The interpreters had been required to turn in their cell phones when they came through the MP checkpoint at the Alamo. The MPs had lost or stolen several of them so I got the policy changed so that they could hold on to them.[88] Cell phones were one of the hottest items on any Afghan's list of 'gotta haves' and to the American MPs a lost cell phone was a bump in the road but to an Afghan it was a tragedy.

The Good Idea Fairies Brief the General

I continued to go to the weekly work sessions of the 'great ideas' group and Major M was really starting to bear down on me to use his slides and say what he wanted me to say about the interpreters. I continued to resist and was able to show him how I was using developmental counseling, right out of the Army leadership manual, to manage the interpreters. Since that fell in line with all the other 'Americanization' of the ANA projects that were being shoved down the Afghan's throats he left me alone, but he did cut my briefing time down to five minutes.

The big day arrived when we were to brief the General. The staff was bursting with self-importance as they demanded our thumb drives so that they could download our Power Point presentations. The atmosphere was almost gleeful in the TOC briefing room as Major M told us where to sit and we watched as he flitted around the room having a word with the chief of staff, and joking with Major Bobby who glared at me whenever our eyes met. I sat in the darkened room with the others waiting for the General. I had spent a little extra time with the Lord Jesus that morning and now I was quietly calling on his name as I waited. I didn't really know what I was going to say when it was my turn to brief but then again I never did. I usually relied on instant inspiration and whatever I saw on the slide.

The General was late and we heard that he was having an attack of dysentery, a somewhat common affliction at KMTC. I had had a few bouts of the 'Kabul Krud' myself since I had been there so I was sympathetic toward the General who was younger than me but still a little long in the tooth for an overseas deployment.

"Attention!" shouted the CSM. We came to our feet as the general shuffled in, wearing a robe and slippers. 'This better be good' I thought as he told us to take our seats. Dragging him off the toilet for our great ideas didn't seem like a 'great idea' but timing was everything and whoever timed this briefing was no prophet.

The S2 introduced the subject and Major M took over from him. Since I was one of the last to brief I had to sit and listen to the same idiotic ideas that these 'professionals' had come up with earlier. Several weeks before, when I had discussed my concerns with the retired LTC at MPRI about what doctrine we should use, he thought that it was a good question: "Are we still in an insurgency or are we conducting stability and support operations?" he had mused. MPRI was supposedly in a lead position in the building of the ANA and he didn't know.

Now I was sitting there wondering how we could build an Army with the officers taught by the French, the sergeants taught by the British, and the soldiers taught by the Americans and hearing how we were going to conduct MOS training for intelligence types using modern U.S. Army doctrine and tactics.

The briefings droned on and I could see that the General was putting a good face on it while he willed his bowels to stay still for a few more minutes. Finally it was my turn. I stood up and introduced myself to the General. I told him how we had thirty-four interpreters at KMTC and though their English skills were uneven we were 'developing' them by counseling and training. The General interrupted me saying, "Well, that's the first time I've heard anybody talking about developing these interpreters.[89]"

The Lord Moves...

I took that to be a good sign and when I finished my spiel about my subject, I turned to the General and said, "And in conclusion sir, KMTC is short twelve interpreter positions..." Out of the right side of my peripheral vision I could see movement as Major Bobby was springing out of his chair. "And I'd like your permission to hire them and get them their back pay."

I barely got these words out of my mouth when Major Bobby broke in with, "Sir ..." The General stood up and said, "Bobby, sit down, I know what you're going to say but I want you to hire those interpreters." That finished my briefing. Major M was glaring at me and Major Bobby was sitting at his desk and staring straight ahead. I didn't see any reason to hang around after the last briefer finished.

The General didn't either and shuffled out of there faster than when he came in. I guess he lost the battle of wills with his bowels and wanted to get out of there before things got ugly where he was sitting. When I got outside I breathed a sigh of relief and began to thank the Lord over and over as I walked back to my tent.

Numbers

The next day I got with JS and told him the good news. I told him to bring the twelve in so we could tell them that they were hired. I would still have to work on getting them their back pay but I wanted them to hear the good news that morning. JS assembled the group and as I watched them come in I counted them. When he told me that they were all here I counted them again. There were only eleven. I counted them again and then I asked JS to count them in case I was losing my mind from stress. He counted eleven.

"Where is the twelfth one?" I asked. "He's in Pakistan," he said. "When did he go," I asked, thinking that he was one of the ones who had gone home for a short visit. "He's always been in Pakistan, he's been waiting for us to hire him;" said JS as if I shouldn't be asking these questions because I must have known all along. Thinking quick-ly, I cut to the chase and told him, "call him right now and tell him to get up here tomorrow morning or we are going to have to find someone else. I just told the General that we needed to hire twelve and I'm not going to get flayed by my friends over at Phoenix because we only have eleven!" After explaining what 'flayed' meant, because JS was always working to improving his English, he got on his cell phone and made the call.

I then turned to the eleven interpreters who had been waiting patiently and told them that God had moved and they were hired. This brought smiles to most of their faces but before they could say anything I forged ahead and told them that my next task was to get them their back pay. This brought bigger smiles to all their faces. After allowing them to thank me for a moment I sent them back to their American partners as full-fledged members of the KMTC Interpreter Corps.

I went upstairs to the head shed and made arrangements with the logistics people to go with them to OMC-Alpha on their next trip so that I could start the process of getting the back pay. This would take me into Kabul for the first time and I would see how the other half lived.

Visit From A Friend

That night I got a visit from Chris, a comrade from SOD who had managed to get a slot on an earlier deployment with the regular army and was now involved in some way with a medical training operation for the ANA in downtown Kabul. He visited with me in my tent and when he saw my copy of 'The Other Side of the Mountain,[90]' a book about the mujahideen tactics in their fight with the Russians during the Afghan Jihad, he asked if he could borrow it and promised to get it autographed by the Minister of the Interior, Ali Ahmad Jalali, the co-author along with Lester Grau, a U.S. Army Soviet specialist.

I had the book in digits on my laptop but I was quite fond of my hard copy and made Chris promise to return it. I still haven't gotten that book back. Chris and his buddy were riding around Kabul with a lot more freedom than we had. It was hard for most of us to get off the base much less ride around here and there in civilian clothes. If I hadn't been so busy with my duties I would have been envious of him. He was in Afghanistan on his third trip when I wrote this and I am happy to say he made it home.

My Interpreter

The next morning I got a call from the MP checkpoint at the Alamo. They had stopped and searched a young man named JW and found a CD in his shoulder bag. When they tried to take it away from him, he grabbed it and ran away. This was the missing twelfth interpreter. He had come from Pakistan after JS called him and told him to get up to Kabul today to be hired. I sent JS out to find him and an hour later he brought JW back to the checkpoint. The MPs wanted to know why he wouldn't let them take the CD and I asked them why they wanted it. They didn't have a good reason for taking it so I told them to drop it and that I would be responsible for him and his CD. I told him to follow me and we went to the interpreter room. When we got there JS formally introduced me to JW, 'the 12th man.'

I asked him why he ran away from the MPs but he wouldn't say anything. Then I asked him if he understood my question and he answered that he understood me but he still refused to say anything. I asked him if he would show me what he had on his CD and he agreed so JS led him to one of the desktops and loaded the CD. JW sat down and took us through the files. They were full of design projects of his that he had created using a drawing program. Some of them were quite intricate and we were impressed. I told JS to assign JW to translation for now and we would assign him to the next American that needed an interpreter. A week later that American turned out to be me.

OMC-Alpha

Two days later I rode with some of the logistics guys into Kabul to the OMC-Alpha resort in the Wazir Akbar Khan neighborhood. This was a collection of nice homes close to the U.S. Embassy that were renting for an enormous amount of money. The houses were now filled with coalition soldiers and civilians that were working under the umbrella name, 'The Office of Military Cooperation-Afghanistan.'

It took a while to find the right office but finally I found my way to the desk of an Army Colonel who looked pretty tired. He had shelves full of big black binders next to his desk and stacks of papers in front of him. He had a large Bible that was open but I couldn't see where. When he finally looked up and acknowledged my presence I mentioned his Bible as an invitation for some fellowship but he ignored that and asked me what I wanted. Bottom line up front, OK, here we go.

I pulled out my interpreter roster and explained to him our dilemma. He looked over my figures and began thumbing through one of his binders looking for a regulation or policy that would guide him on how to deal with this problem. "I don't know what we can do about this," he said finally, expecting I guess for me to agree with him and leave. "Well sir, I think you better dig a little deeper in those regs because we have to give these guys their back pay, it is a moral imperative," a favorite phrase I had been using a lot lately, every time I encountered someone like this who thought the Afghans didn't matter.

He looked at me like I was crazy. Then he started thumbing through another binder until he found something and asked me to give him more details about how the problem had started and who was responsible. I gave him the whole story with names and ranks. When he heard that the Vermont commander was a full colonel he thumbed through a few more pages and said, "I think we can justify $3,700.00, what do you think?"

I thought about it for a minute and said "Thank you sir, that would be perfect." Once again the Lord had come through with more than I could ask or think. I gave him the information he needed to direct the funds, thanked him and left. I went to the PX and sat around while the others completed their work, watching the OMC-Alpha people flittering here and there in their little paradise in downtown Kabul.

Care Packages Arrive

I got an email from the woman in Dallas telling me that the boots and jackets were on the way. I could expect them in two weeks, wrote the woman and I could pick them up at the parcel delivery company's office in downtown Kabul. I got permission from the CO to organize a small convoy for the pickup. About ten days later I got an email from the delivery company telling me that the packages had arrived. I contacted the people who had agreed to go downtown with me and we left KMTC after lunch and drove into Kabul. I hadn't been downtown since the trip to OMC-Alpha so I rubbernecked and photographed or videoed everything.

The whole scene reminded me of places I had been in Central and South America. The crowded streets with cars competing with horse carts and the open air butcher shops and fruit and vegetable shops were just like places in Nicaragua and Colombia. The camels served to remind me that I was a long way from those places. Large military vehicles of the German Army careened through traffic with German soldiers behind machine guns that looked a lot like MG-42s. The Iron Crosses on the sides of these behemoths caused me to fantasize that the Huns had risen from their own ashes and were now on the move, again.

The traffic in downtown Kabul was intense and it seemed that the only traffic law that was being observed was that the larger vehicle with the most potential firepower had the right-of-way. Everything else depended on your skill and aggressiveness. The American soldiers would demand right-of-way by aiming their rifles or pistols out the window and making the Afghans yield. I hated this practice because if we weren't an occupying force, then we needed to stop acting like one.

The Afghans had been told that we had been asked to assist the Northern Alliance in ridding the country of Al Qaeda and the Taliban who sheltered them. That was why most of them were happy to see us. Holding guns in their faces wasn't going to enhance that perception. Some of the other soldiers from RI decided that I was an Afghan lover and didn't want to go anywhere with me in a vehicle because I refused to point my pistol at people I didn't need to shoot.

We took one of the interpreters from the logistics section and he directed us to the office. When we got there I gave them the package information from the email and soon we had five large boxes. I signed for them and we loaded them on the truck. When we got back to KMTC I put JW in charge of sorting out the boots, jackets and some other clothing and putting each set into a separate plastic bag with an interpreters name on each bag. When he was ready we gave a bag to each interpreter.

The next day I waited by the entrance to the Alamo to see the interpreters in their new jackets and boots. As they came in I was disappointed to see that everyone was still wearing sandals. A few of them were wearing the new jackets. Most of them were professional sports team jackets with bright logos. I couldn't believe that none of the boots fit because we had gone to a lot of trouble to make sure that we had the correct sizes.

When JS showed up in sandals I took him aside and asked why he wasn't wearing his boots. He looked a little nervous and told me that he had given them to his brother. I told him that that was kind of him and that I would get him another pair. He smiled and thanked me but said that it wasn't important. When I reminded him that it was still cold and would be for several more months he let me in on the big secret. "We always wear sandals here," he said with a shy smile. "You mean you would rather have cold feet than wear those warm boots?" I asked. "The people would laugh at us if we wore them," he said a little sadly, "It's the style."

Matt

Matt, the KBR kitchen worker and I had started meeting on Friday afternoons at the front gate. At first it was for English lessons and we found a quiet place in the large building that was occupied by the Camp Phoenix interpreters. They had a large room at the end of the building near the front gate and we found a small room at the other end. There were a few old mattresses lying around and we used them to make seats on the floor while we leaned against the wall.

Matt had a notebook that he was using to add English words to his vocabulary and he had a small Dari/English dictionary that he used to help him. We would spend about an hour talking about whatever was on his mind. By using the dictionary, hand gestures and having the belief that we were understanding each other because we wanted to, we passed a pleasant time each Friday. I emailed my son and my wife and asked them to send me a guitar so I could sing some spiritual songs to him. I also asked my wife to find a Dari New Testament so I could show it to Matt if he was interested.

For I also am a man under authority, having soldiers under me. And I say to this one, Go, and he goes; and to another, Come, and he comes; and to my slave, Do this, and he does it.
Matthew 8:9

Chapter Five: March 2004

A New Office

I came to work one morning and after the staff meeting Major Shawn the S3 told me to come to his office. "I'm moving Major Mike upstairs and I want you to occupy his office." Whoa, I thought, Major Mike is not going to be happy about this. He had been in the office right next to my desk and kept the door shut most of the time. I could usually hear him talking to his wife on the internet phone when I was sitting at my desk and so I was pretty sure that this was some workload issue between the officers.

I was very happy to get the office though. I had started a project using an intel preparation of the battlefield format to do a study of the KMTC area and I needed a space to put up a large map for determining possible threat locations. I would use the map to designate likely enemy observation points based on line of sight analysis.

When I got downstairs Major Mike was packing his things in preparation for his move upstairs, so I went outside and crossed the courtyard and went up to the interpreter room. They were all buzzing away in Dari and Pashtu. We had several computer work stations and there were groups of six or seven men in a cluster around each while one of them showed the others how to do something on the computer.

Interpreter Activities

I remembered the computer training we were doing at CD to transform the cargo inspectors into intel analysts. I decided then that I would contact Air Marshall Jim and see if he could send me a copy of the software that taught you how to use all the components of MS Office. I told JS about it and he agreed that most of the interpreters wanted to learn computer skills and would appreciate some in-house training when they didn't have interpreting or translation duties.

The next thing we talked about was the language training that was being conducted for the workers at KMTC. MSG Bill had started literacy classes and some of the interpreters were teaching their fellow Afghans how to read and write Dari. Every day at lunchtime, about thirty men would come to a classroom at the Alamo and spend forty-five minutes studying their language. The years of war had destroyed the education system of Afghanistan but these men were willing to give up the only break they had during their day of hard work so that they could learn to read and write. This told me a lot about the Afghan character.

I wasn't able to pay these interpreters like Bill had done. He was a one man dynamo, working money raising deals all over KMTC, Camp Phoenix, and Camp Blackhorse in Pol-i-Charki. I didn't want to shut these classes down. The worker's need for education and the fact that the ANA General liked the classes weighed heavily against that, so I talked to JS about the problem which came down to money.

He had been telling me about how things went in the early days of the operation at KMTC when the 3rd Special Forces Group had a FOB at KMTC. They were the ones who trained the first Kandaks along with the French and British. The interpreters were well taken care of by the SF. JS described the SF as being laid back and generous. When there wasn't a need for the interpreters the SF left them alone. They were paid better then and the SF guys had a lot of money to spend on things like TVs and VCRs for the interpreters to watch movies when they were off-duty. When Camp Phoenix was established and the SF moved on to more exciting work elsewhere, the lead interpreters took the TVs and VCRs to Camp Phoenix. Now the interpreters at KMTC were being used for both interpreting and translation with less money and fewer benefits.

I told him that the freewheeling days of the SF way of doing things were over. We had to take care of perceptions at KMTC because some of the ANA officers were jealous because they were being paid less than the interpreters. This led to a small amount of tension that I didn't want to get any bigger. I asked him if he thought the men would want another TV and VCR in the break room and he assured me that they would.

In my less than subtle way I linked these things with the need to continue the literacy classes without pay. I emphasized the fact that the ANA General liked them and since JS was close to this General he agreed with my assessment. Since I had managed to get the twelve interpreters hired and get their back pay he was a lot more receptive to my recommendations and agreed to keep the classes going for now.

Latrine Law

We were outside during part of this conversation because JS needed to use the latrine. That's when I became aware that the Afghan interpreters were all using a 'Port-a-Potty' that was behind their building. They washed their hands using a water faucet nearby. It was still winter and I thought that they were probably pretty cold doing their business out there. We had just gotten a 'Force Provider' latrine installed at the Alamo but the Afghans were prohibited from using it. I decided to change that and after JS and I finished talking I went inside and wrote a memo that allowed the Afghans to use our latrine.

I took the memo to LTC Randy Davis, the Garrison Commander, and asked him to sign it. I explained my rational; he agreed and signed the paper. I went back to JS and told him that the interpreters were now authorized to use the latrine. The Afghans seemed to be happy with the idea of having a heated toilet so I went on my way, believing that I had advanced the cause of Afghan/American friendship another yard.

The Staff Time Blues

The next morning I went to the staff meeting. I saw Major Mike and he didn't appear to be very happy. I went over to him and told him that I had nothing to do with his being moved out of his office which was now mine. He thanked me but I could tell he was disappointed about having to move upstairs to an office just outside the CO's office.

The staff had to deal with the CO on an instantaneous basis every day and they were all looking pretty ragged because it was a stressful job. We were running a large installation with more than 5,000 trainees and the staff of a foreign army that was totally dependent on us. We had multi-million dollar construction projects for which we took bids, let contracts, and quality controlled the construction. Others were responsible for procuring and issuing the weapons, ammunition, and equipment for the trainees along with food, medicine, beds, blankets, toilet paper, etc., etc., etc.

This was a major undertaking and our people, most of whom had never done anything like this before, were making it happen under the leadership of the CO. He seemed to be a dour and somewhat mean little man but he was making us complete the mission to acceptable standards by the force of his will and for that I had to respect him.

One of the primary responsibilities of the commander, which he did, was to create and foster a unit climate that makes it difficult for someone to give in to their weaknesses. That doesn't mean that I liked him and it didn't seem that he liked me or most of the others in our unit. As we settled into this task I realized, during the fleeting moments that I gave to such things, that these were not going to be happy times fondly remembered in years to come.

The Modern War—Connected

After the staff meeting I went back to my work area and moved into my new office. I contacted SSG Carl and asked him to hook me up on the network. He not only hooked me up but he helped me get connected to an on-line telephone system that cost very little. He also uploaded Falcon View, the latest and greatest mapping and mission planning software used by SF. He was a high speed skydiver and he had friends in the special operations realm.

When I got my laptop back from him I was authorized to use the high speed (literally) internet system that was the envy of the people back at Camp Phoenix. I was sure that they would figure out that we had a much better connection than they did and take it away from us, but somehow SSG Carl defended our system from all attempts to get it and we stayed on line until we went home.

After I checked in with JS and found that all was well with the interpreters, I went back to my new office, closed the door, and conducted some top secret 'intel' activities; I called my wife on the internet phone. After a few rings she answered with a sleepy voice and I remembered that we were almost twelve hours apart and I had awakened her from a deep sleep. She rallied quickly though and we spent twenty minutes bring each other up to date.

I kept praising the Lord for the way he hired the twelve interpreters and got them their back pay in the face of all the opposition from the S2 of the 45th Brigade. Then she told me that my son was not doing well in school. He had gotten into on-line video games dedicated to fighting terrorists in a worldwide coalition of video game addicts and it was leaving him so tired after firefights that lasted several hours he didn't have any energy for homework. I told her I would email him and I told her to thank him for his attempts to defeat our enemies and to tell him that if he didn't graduate from high school I was going to declare him a terrorist when I got home.

Advice and Consent

Major Mike came to my office one morning and invited me to a meeting with the ANA operations officer. He said that there was a new ANA officer who had been assigned to be the S2 for KMTC and it would be my job to mentor him. I was busy that morning and begged off. He told me he would get back with me later. For the next couple of days I went to the staff meeting and nothing was said about any additional duties. I figured that if there really was a new S2, and the CO wanted me to get involved he would have said something. Then Major Shawn, the S3, told me that I should go with Major Mike and meet the new ANA S2. When I asked him why I wasn't getting my marching orders from the CO he just laughed and told me to go to the next meeting with Major Mike.

I went to the next KMTC security meeting with Major Mike and SFC Rod, the leader of the Ninja Turtles, our security advisory team. The meeting was in the office of the KMTC security officer, a Major NS. He was about six feet tall and he looked like a football player. His office was very small but there were several chairs and he invited us to sit down. A sergeant came in and began pouring tea for us and I was introduced to Major NS.

After we finished the tea, Major NS told us that the meeting would be in Colonel Z's office so we followed him to the Colonel's office. There was an Afghan Army officer sitting at the table who I hadn't seen before and I guessed that he was my new counterpart. We sat down at the table with our interpreters and Major Mike introduced me to Major TN. He was about 5'4". He had dark hair streaked with gray, a mustache and was clean shaven. We had more tea and then the meeting began.

Major TN shook my hand and we sat drinking tea while Major Mike and the security captain discussed some issues. Since I didn't speak Dari and Major TN didn't speak English, I suddenly realized that I needed my own interpreter. JW, the mysterious 12th man would have to be the one. Major Mike's interpreter helped me establish that I would meet Major TN every day at the place of his choosing. He told me to meet him in the office that we were in the next day after lunch. Major Mike seemed happy that he had gotten me linked up with the new S2 but couldn't tell me anything about what was expected from me with regards to an S2 at KMTC.

My Interpreter

When I got back to the Alamo I told JS that I wanted JW to be my interpreter. When I asked him if he knew anything about the new S2, he said no. Then I walked over to where JW was working on a computer. There was a group of interpreters standing behind him, watching him create a colorful design on the screen of the computer monitor. I waited until he stopped for a moment before interrupting him and asking him to come with me. The other interpreters were like the parting of the Red Sea waters as they made way for him to come to where I was standing.

I told him to follow me and we went down the stairs and crossed the courtyard. He followed me to my office and when he was inside I closed the door and told him to sit down. He was nervous but tried not to show it. He had no idea what this was about so I started by asking him again why he ran away from the MPs the first day. He had loosened up a little since then and now he told me that he was afraid they were going to take away his CD which contained his portfolio of designs. I asked him if he had any problems with the security people at Camp Phoenix and he said that they had interviewed him and passed him for working at KMTC. I knew this but I wanted to feel comfortable with him and now I believed his explanation about the incident with the MPs after seeing how much he was into computer graphic design.

I told him that I wanted him to be my interpreter. That brought a smile to his face and so we began a close relationship that has lasted to this day. We talked about our backgrounds; he was a student who had hoped to go to medical school and had been living in Pakistan with his mother and sister. I knew from JS that the only people who went to medical school were the ones who scored the highest on the national examination that was given to the high school students after their graduation. JW was smart and his English was very good. I told him about my background as a high school dropout and that I was a Rhodes Scholar. He seemed startled by the contradiction so I told them that I got that scholarship when they told me to hit the 'road.' He got it immediately and laughed so I knew we would get along.

Then I explained to him about our need to think the same thing when he was interpreting my words to the Afghans. He was to help me if I was being impolite or foolish but I wanted him to stick with me and not run his own program like some of the interpreters would do. As long as he asked me for clarification before he spoke something that he didn't understand we would be fine. I told him that I wanted to look at the person I was talking to so he would have to sit close to me in order to hear what I was saying. We talked about these kinds of details as we got to know each other and I was glad we had that time because the next day we would be operating with the ANA S2.

The Needs of My Counterpart

The next day we had our first meeting with Major TN. He didn't have an office yet so we met in the security officer's office. After we drank some tea and exchanged the mandatory pleasantries regarding our health and the health of our families, we got down to his business. "Before I can begin my work I will need some things," he said. I asked him to continue and that opened a floodgate. He needed many things to include an office, a vehicle, some computers, guns, personnel and equipment. I waited until he finished and then told him that we would work on it. He didn't look very happy with my reply and said something to JW

JW looked embarrassed and hesitated when I asked him for the meaning of his reply. Finally he said, "He wants to know why you didn't write these things down." Oops. I was caught and told him that I had forgotten my writing material but I would remember what he told me and would write it down when I got back to my office. He still wasn't looking very happy with this but I had to leave and after we agreed on a time for the next day we left.

JW said that Major TN was not happy because I didn't write down everything he said so when we got back to my office I had JW help me make the list from both our memories. JW and I thought it was funny the way Major NS said that he couldn't begin his work until he got the things on the list. The way my wish lists had gone so far Major TN would be idle for a while.

An Advisor Gets Some Advice

The next day when we returned to the Five Story Major TN was waiting for us. After the tea and sympathy we got down to business. He made a long statement to JW; the gist of it was that as an Intelligence Officer, I had to have an intelligence notebook. He smiled while reaching into his briefcase. I was waiting for a gun or a hand grenade because we didn't know each other very well at that point, but he produced a small notebook which he handed to me. I have since read about the Russian KGB practice of special notebooks for intelligence officers so I guess he had been a member of the DRA's intelligence service (KHAD) during the Jihad.

I took the notebook; it was brand new and I thanked him profusely. He then began speaking to me and JW translated his instructions about how to use the notebook: make sure I recorded the dates at the top of each page; make sure I stored it in a safe when I wasn't using it, etc. I was getting annoyed but I realized that our relationship had everything to do with how he perceived me and since I had thoughtlessly failed to write down his list of needs he had earned the right to school his 'advisor' on basic intel matters. I bit my tongue and learned a lesson.

When we finished the notebook lesson, he proceeded to repeat his wish list of yesterday and this time I wrote everything in my new notebook. He added a few items and the list now filled several pages. When we finished with that we talked a little about our backgrounds. I told him that I had been a fighter in Vietnam when I was a young man. He told me that he had heard that the Americans had lost that war to the communists.

I detected a little pride in his delivery so I let him enjoy my nation's humiliation by admitting that we had indeed lost that war but I hoped that we could work together to make sure that the Afghans didn't lose this war to the Taliban. He agreed. This was how we spent the next week; each day he would go over the list of what he needed before he could begin his work and I would follow along with my intelligence notebook. After the meetings I began visiting the logistics guys at the Alamo for guidance on how to get the wish list into the pipeline.

Linguist

Now that I had my own interpreter, I decided to try and learn a little Dari, the main language being spoken at KMTC. I had been wanting to give a blessing to the young ANA soldiers that I met from time to time around the camp. I asked JW how to say, 'God bless you.' He thought for a moment and then said, "Hoda Bobachet." I practiced saying it, letting him correct me until I got it right. Then, the next day, when I came across a platoon of young recruits taking a break, I walked up to them and brightly proclaimed, "Hoda Bobachet!"

As one, they stepped back away from me and began looking at each other with expressions that usually indicated alarm. Believing that I had mispronounced the blessing, I tried again, even louder. They shrank back even further in alarm. Since I was by myself and had no way of explaining myself I waved to them with my biggest smile and returned to the Alamo to find JW. We had a meeting with the ANA operations officer and I didn't want to be late.

On our way to the Five Story I asked him to listen to my pronunciation of the blessing and he said it was perfect. When we arrived at the Five Story I decided to try it again and greeted the Afghan colonel with, "Hoda Bobachet!" He looked me right in the eye and then turned to JW and said something in Dari. JW translated, “He asked you why you said, may God forgive him, because he's not dead yet.” That freaked me out, remembering the look on the young Afghan soldiers’ faces. Later, after talking to JS, I found out that in Pakistan, Hoda Bobachet means 'God bless you'; in Afghanistan it's what you say for a dead person at their funeral. I guess the soldiers thought that I knew they were being sent to some bad place when their training was finished. While that was indeed likely, what we had here was a failure to communicate.

Intel—'The Secret World'

I was continually bothered by the lack of guidance from the command regarding Major TN's 'work.' No one was saying anything about what his duties were. I began to conduct an investigation to determine who was doing what at KMTC with regard to security. The security mentoring guys, the 'Ninja Turtles' were riding around the countryside near KMTC so I asked them what intel they were getting and from whom. I assumed they were operating as part of a well-oiled machine being run by Camp Phoenix. SFC Rod said that they pretty much did what they wanted to do because the people at Camp Phoenix didn't have much to offer as far as current threat information. Since I was talking to the same people every morning before sunrise I knew that to be true.

The Intel MOS course was being taught at PEC so the only thing I could see the S2 doing here was establishing an SOP for providing intel to the security forces based on an intel estimate with a counterintelligence section that would guard against enemy penetration of the ANA and KMTC. Every time I raised the subject with Major NS, he would remind me of the list of things he needed before he could begin his work.

The Combined Joint Coalition Task Force

Camp Phoenix, our home away from work, was an interesting place. Instead of eating with the Americans I enjoyed sitting down at tables in the DFAC that were occupied by our allies. I spent a pleasant time, for example, with a captain from Romania who told me that with the extra money he was getting from this deployment he was going to buy a small house and get married. The word had gotten around that I was an intel guy so most of our allies were a little standoffish when I sat down but eventually someone at the table would try to engage me in conversation. Of course, unless they spoke Spanish or English, it was not going to be a long conversation on my end but most of our coalition partners spoke some English and we were able to exchange pleasantries.

I tried to make friends with some of the French soldiers but they had their own dining facility and were still smarting from the American barrage of anti-French sentiment on the internet resulting from their government's refusal to participate in our invasion of Iraq. One of the things floating around was a parody of 'Soldier of Fortune' magazine called "Soldiers of Surrender" with stories about white flags, 'a new technology that could save your life' and the latest results of 'drop tests' on weapons systems. It was pretty funny but not conducive to coalition building. The French soldiers I met were like soldiers from any army. The young ones wanted to get in the fight and prove their manhood. The older ones were proud of their past service in places most of us had never heard of. There was one thing that both the French and American soldiers had in common which no one was mentioning—both our countries had been defeated in Vietnam.

Good Morning—Afghanistan

Every morning after I had finished my intel update and breakfast, I returned to my tent and rested before it was time to drive to KMTC. The 45th Brigade had put a large screen TV in every tent and the Armed Forces television station, which was the only station we could get, was showing some guy named 'Doctor Phil' each morning at that time.

It reminded me of 'LT Frenchie' in the movie, "Good Morning, Vietnam" in that it was the most inane programing I could imagine for troops in a combat zone. Maybe it was designed to complement the anti-depressant medication that was beginning to be prescribed for all complaints the way the 'all-purpose capsule,' Darvon, and Motrin were prescribed in the past. I learned quickly to tune out whatever 'Dr. Phil' was talking about and to this day I can tell you very little about him other than that he looked like a pimp and talked like a con man.

One Step Up-Two Steps Back

I got a call from LTC Davis and went to his office. He had received some complaints from MPRI about the Afghans using our latrine. One of the kitchen helpers had stood on the toilet seat with muddy shoes and left an indicator of his poor aim on the back of the seat. There had been such uproar that LTC Davis was forced to rescind the policy allowing the interpreters to use our facilities. I felt sick to my stomach when I went to JS and told him that the interpreters would have to go back to using the single port-a-potty and cold water faucet.

This was not the first time we had problems with shit at KMTC. The city dwelling Afghans used a different type of toilet, one that you straddled. The country dwellers had less sophisticated methods. When the first port-a-potties were brought to KMTC, the soldiers thought the blue water was for ablutions and the legend of the 'Blue Men' at KMTC entered the war story collection. After that notion was disabused came the problem of rocks.

The ANA supply officers were reluctant to put toilet paper in the port-a-potties because they couldn't believe they would ever get more to replace what was used. They either hoarded it or sold it. The soldiers reverted to the country method and wiped with rocks. When the SST (Shit Sucking Truck) came to upload the shit, the rocks were sucked into the pump and destroyed it.

The problem of toilet use was continuous. After the sewage system was restored at KMTC we would still have continual problems with rocks in the pipes in the barracks. All this crap resulted in a daily Power Point slide in the staff meeting appraising the colonel on the state of fecal matter disposal at KMTC.[91]

Badges, We Don't Need No Stinkin Badges!

When I was introduced to Major NS I also met the colonel in charge of operations. He was working on a plan to issue ID badges to all the workers at KMTC. The threat of suicide bombers was increasing and the ANA security force was trying to come up with some procedures that could prevent one from coming on the base. Our operations people and the security mentoring team, the 'Ninja Turtles' were not interested in helping the colonel who had few resources of his own and he was desperate.

We had a few pleasant conversations while I was getting started with Major NS so he asked me if I would help him. I already had plenty to do and I was a little pissed at the others who felt it was much more important to daily explore the far reaches of KMTC and beyond than to help their counterparts with the badge project which was being driven by General A. After trying to work my way out of this additional duty I gave in and we entered into serious discussions regarding what was needed and what he wanted. This led to some interesting discussions about some fundamental problems that the Afghans were facing.

First, he wanted badges for each type of worker. On the badge he wanted their name and nationality. His first inclination was to use the person's ethnic ties, like Tajik, Uzbek, Hazara, or Pashtun. He asked me what I thought. I had just finished reading a draft of the new Afghan constitution and told him that the constitution called all Afghan citizens, Afghans; he liked that and agreed that the nationality block would say, Afghan, or Pakistani, Indian, or whatever.

JW was taking notes of his own because as a graphics illustrator he decided he was going to design the badges even before I asked him. He entered into the conversation on his own behalf and started discussing the badge design with the colonel in Dari. Soon everything was humming and JW ended up designing all the badges for the civilian workers at KMTC.

I sat with him later and we worked on the kitchen workers badge. I suggested a crossed spoon and fork above a plate. He liked it and when he showed it to the colonel, he approved it with a smile. We had just completed a new large mess-hall for the ANA. I ate lunch there occasionally with my interpreters. The menu was the same every day, a large serving of rice over a piece of meat or chicken and a large piece of delicious bread. The daily crowd made it a tempting target and I hoped that these badges would help keep things safe there.

Alas, this was not to be. A few months later a fight broke out there and someone yelled, "Taliban!" The troops inside panicked and tried to get out. The sergeants outside started throwing rocks at them to keep them in. All the windows were broken and the furniture was destroyed. Witnesses said it looked like a food fight in the movies.[92]

The Great Game I

Beside the occasional request for help I was content to supervise my interpreters while trying to figure out how to get the things my new counterpart needed so that he could 'begin his work,' an activity that was still undefined. I had started working on an intel estimate of the situation at KMTC; if there was anything like this done by the Vermont intel officer it was still locked in the safe and no one had a key.

JW and I made one of our daily visits to Major NS. I made sure I had my notebook and pen before I left my office. It was a good thing because today I would need it. When we got to the security office Major NS was waiting for us as usual, and there was an Afghan civilian sitting in one of the chairs when we walked in. We did the usual greetings and then Major NS asked the security officer to leave and he introduced us to a man I'll call Ishmael.

Ishmael was a walk-in. In the past he had been a member of KHAD, the Afghan Communist government's intelligence agency. He came from the Pakistan border area and wanted to get back in the 'great game.' The ANA officer he contacted had sent him to Major NS and that was how we were hearing his pitch. He had information on a particular person who had supposedly been a big shot in the Taliban before the U.S. invasion. According to Ishmael, this man, who we'll call Timor, was now a colonel in the ISI and was working in the tribal areas along the border recruiting men for Jihad in Afghanistan.

Ishmael had information that seemed credible in itself and even though I had no way to judge his reliability I thought that someone should hear it. I asked him for all the details and wrote them in my notebook under the approving eyes of my counterpart. Of course the matter of money finally came up. I told Ishmael and Major NS that I had neither money nor the authority to act on this information but that I was going to try and connect Ishmael with some people who did.

Ishmael was going to stay with Major NS for a few days so after arranging to meet them the next day I returned to my office and wrote a report which I printed. When I returned to Camp Phoenix I took it to the CI team. They appeared to be interested and after reminding me once again that they were the only authorized intel collecting people in my world, they told me to send Ishmael to Camp Phoenix the next day for an interview. I asked them to come to KMTC to protect Ishmael's cover but they insisted that they knew what they were doing and for him to come to Phoenix.

The next day I met with Major NS and Ishmael to give him instructions on where to go and who to ask for. I cautioned Ishmael that he should leave the money part out of the conversation until he was sure he was talking to someone who wanted to act on his information. I wasn't offended by the money part and wished that I could work Ishmael myself for the chance to strike a blow against our enemies.

Two days later Ishmael was back with a tale of woe. The CI team was only interested in whether he had information regarding direct threats to Camp Phoenix. Ishmael tried to explain that Timor was recruiting men to attack coalition forces all over Afghanistan including Camp Phoenix, but the CI guys told him to hit the road. Major NS was indignant and Ishmael was dismayed. After assuring them that I would continue to try and find someone to listen to Ishmael's information, I asked them to meet with me the next day.

This was ridiculous. I went to the CI team's tent as soon as I got back to camp. They had lost my report, thrown it away was more likely and they were on the defensive about their actions. I knew these guys didn't have any experience with this kind of work so I tried to patiently walk them through the concept of researching the information that Ishmael had given us to see if there was any connection to what was already known about the subject. They blew me off pretty quickly, once again reminding me that they were the only authorized people to conduct any kind of intel collection operations.

The next morning I went to the Brigade S2 shop for the morning intel dump. I began to feel like I was going to the latrine to pick up someone else's 'dump 'for all the good it was doing us. The night shift LT had always treated me well so I asked him if he could tell me if there was any information about Timor on the SIPRNET, a classified internet system. He ran a quick check and told me that Timor was a HVT, a high value target, with a $100,000 reward for his capture. It had taken him 30 seconds to find this information; this made my anger with the CI team grow hot.

When I met with Major NS and Ishmael that afternoon, I apologized for the CI team's less than warm appreciation for Ishmael's information. I told them that I would be trying to find someone to listen to his information. They listened but it was hard for them to imagine that the Americans could be so incompetent regarding this kind of information. They probably suspected that I was either the stupidest guy in the intel business or that I was working some scheme that would enhance my possibilities for promotion and a cut of the reward they knew must be out there for someone of Timor's stature. Osama bin Laden had a twenty-five million dollar reward on his head and there were a lot of people in Afghanistan trying to find him.

After I determined that Ishmael wanted to continue to work with us, we discussed methods to contact him where he lived so we could tell him when we needed him to come back. When the tradecraft procedures were established, I promised Ishmael that I would do what I could to get his information to the right hands. Before we left we hugged each other following the custom of the Pashtuns. On the way back to the Alamo JW asked me why I hadn't searched Ishmael for bombs; suicide bombing was starting to be popular in Afghanistan. He was right and I told him that I would do better next time.

Problem Solving 101

The word started to get around the Five Story that there was an American who would help with a problem. The ANA officers began to greet me in the halls and out on the parade ground. I had begun to salute the officers because we saluted the coalition officers and I thought that the ANA officers would feel better about us if I did. This never caught on with my fellow Americans as far as I know but I enjoyed surprising the Afghans with my best 'Old Guard' style of salute.

Thanks to my 'notoriety' I was stopped occasionally and pulled into an office. The ANA officer would give JW an earful and then he would tell me what was up. One day we were led into the office of a logistics officer in the Five Story. He had several Afghans in traditional clothing. They all began to speak to JW at once and it took a few minutes for the ANA officer to quiet them down. Then he gave JW some more information which he eventually relayed to me.

The men ran a transportation company. They had been contracted by the Vermont guys to haul away some derelict tanks from the center of KMTC and they had never been paid. One of them continually waved a piece of paper at me while the ANA officer continued to explain these things to JW who then relayed them to me. JW grabbed the paper at one point and handed it to me.

It was a work order for $7000.00. It was signed by one of the LTCs from Vermont. The ANA logistics officer told me that I was an American who would help them. I didn't argue; I took the paper and told them I would get back with them. I took it to our logistics people and two weeks later the Afghans were paid. The ANA logistics officer was very gracious to me after that and I chalked that one up in the win column for the battle for the hearts and minds of the Afghans.

The OGA

There was a certain other U.S. government intelligence agency, the 'OGA" or other government agency, which had a 'secret' base nearby. We had some dealings with them in the past when stray bullets from their firing range landed in people who were using our range. Several Afghans were wounded and we felt that the OGA would cooperate with us by changing the direction of their target practice but they didn't. Our Garrison Commander, LTC Randy Davis visited their compound a half a dozen times to no avail. Our people appealed to OMC-A, then the embassy, the highest military/civilian authority in Afghanistan but nothing changed. They were a law unto themselves.[93]

A few weeks before, one of our logistics officers, the major who had arranged our C-17 transportation into the country, had mentioned to me that he was in contact with some people over at the OGA compound. I went to him and asked if he could hook me up with someone over there so I could pass some information. A couple of guys showed up the next day and asked me what I had. I gave them the report and let them read it. They were interested. I had heard that, among other things, the secret base was being used to train a special force of Afghans that would go after the HVTs when they were located. They asked me to arrange a meeting with Ishmael and I said I would but I needed some assurances that they would pay him for his troubles since he would have to come a long way. They absolutely committed to pay the man for any credible information and left.

When I saw Major NS at our next meeting I told him we were on and to activate the recall procedure we had established. Two days later Ishmael was sitting in his office after a long journey from his home town. The OGA guys showed up after a phone call and I introduced them to Ishmael. They asked if they could borrow one of my interpreters. JW was on a home visit so I gave them one of the better English speakers who wasn't doing anything at the moment. They all sat in the Suburban they were driving and talked for about 30 minutes. When they were done, the interpreter came to my office and got me. I went out into the courtyard in time to see the Suburban drive away.

I walked over to Ishmael and asked him through the interpreter what had happened. He said that they asked him a lot of questions; then told him to get out and they drove off. Now I was really frustrated. I still thought that Ishmael's information had the potential to be useful in our attempt to stop the Taliban insurgency that was bound to grow if we didn't stay on them with an intensive human intelligence effort, 'spies,' that could provide 'actionable' information.

I took Ishmael back to my office. I apologized for what had happened. If he still wanted me to try and connect him with some intel people, I would, but it was going to take more time. He did, so I looked in my wallet and took out the only money I had, a twenty dollar bill, and gave it to him. He gave me a big smile and a hug. He must have thought that he had found the money river and although it was only a trickle he was encouraged. We agreed to follow the same procedures for recall and I took him back to Major NS's office.

I explained what had taken place with the OGA and I could see that he was disappointed. If we had a success with this he would benefit by a share in the reward money that I was sure he would have arranged with Ishmael and it would have made him look good with his own chain of command that had no love for the Taliban. We all agreed that I would continue trying to find the right people and that we would follow the same procedures when I found someone.

The OOGA

About a week later, after the staff meeting, I walked over to the Five Story in hopes of finding Major NS. I crossed the parade ground near the HQ building and met several of the interpreters who were watching some ANA units doing dismounted drill. We talked for a while and as I turned to leave, I saw a couple of bearded Americans wearing Royal Robbins type gear, the fashion statement of this war, so I figured they were some kind of spies. One of them seemed familiar so I looked at him carefully out of the corner of my eye. He noticed my attention and smiled. It was Steve, an old buddy from the police and SOD who I hadn't seen for 10 years.

"Dude," he said, "what are you doing here?" I laughed and asked him the same thing. We stepped away from the others and he told me he was working for an unnamed agency doing intel. "You're working for the OGA?" He hadn't heard that one so after I explained it he laughed and said that he was working for the OOGA, the 'other other government agency.'

Maybe he could help me, I thought. He was here on a scrounging mission and needed coffee. That didn't surprise me because when we were cops he lived on caffeine. I promised to give him some of the coffee we had stored at the Alamo in exchange for a few minutes of his time. I sent JW up to Major NS and asked him to stand by while I walked with Steve to the Alamo.

On the way I poured out my lament about the intel situation at the 45th Brigade and the rest of Afghanistan. He was always upholding the established order of things and was not very sympathetic to my complaints but when we got to the Alamo and I showed him the large stash of coffee he agreed to listen to what I had to say about Ishmael's information regarding the HVT, Timor.

We went to my office and I let him read my report. When he was done I pitched the concept that someone who was working the cross-border area should evaluate this and if it complemented what they already knew then maybe Ishmael would be useful as a spy. He agreed and told me that he would talk with his people and get back with me by email. I was to make sure that Ishmael would be available when he came back. I told him to give me a two day advanced warning and helped him load some coffee and food into his SUV.

The Estimate

I continued to work on the Intel Estimate for KMTC and as I developed my thoughts I began working on a decision brief that Major NS and I would give to the General commanding KMTC. I hadn't gotten any guidance from the TAG command group and Major NS's daily litany of things he needed before he could 'start his work' was starting to sound like fingernails across a chalkboard. The only way I could shut off the flow of frustration was to wait until he had to take a deep breath and jump in with questions requiring his professional opinion. He was always happy to school me on the methods of a Soviet trained intel officer and together we hammered out our proposals to the General.

General A was a one man show as far as I could tell. He had a large staff but no one did anything without his approval. I saw many an officer in his office getting authorization for the issuing of toilet paper or some other innocuous staff functions. Decentralization was not part of the ANA doctrine so if we wanted some things like office space and personnel, we would have to ask him.

As we developed our SOP and briefing I was having JW translate the SOP verbiage into Dari and he also worked on the Power Point briefing. Working with JW was fun because even though he was young he had a cynical view of bureaucratic life. Eventually I found that most of the Afghans saw things the same way due to the prevailing thought in that part of the world that everything was a conspiracy and nothing was what it appeared to be, a good mindset for intel work.

The Doctors' Plot

I was working at my desk one morning when CPT M the medical officer asked me to go with him to a meeting with the medical doctor translators. These men were all medical school graduates and were translating American military medical classes into Dari to be used by the ANA in the training of its medics. They were working as translators because they could make more money doing that than by practicing medicine.

They had complained that the Afghan doctor who had hired them was extorting money from them as a kind of 'kick back.' He held the contract with OMC-Alpha and when they got paid he kept twenty dollars from each of them each month. Doctor A was a Tajik and all the other doctors were Pashtun.

CPT M asked me to say something so I told Doctor A that the U.S. didn't allow this kind of business practice. I said that the best way to resolve this was to give back the money and make peace with each other. Doctor A agreed to do so and produced a one hundred dollar bill. I gave him five $20 bills which he distributed to the doctors. I was happy with my great negotiating skills and after shaking hands with all the doctors I left, noticing that none of the doctors seemed that happy about the outcome. The next day the Counterintelligence team from the 45th Brigade showed up and arrested the five Pashtun doctors for being Taliban spies.

Doctor A, either on his own or with advice from someone else, went to Camp Phoenix and denounced the five Pashtun doctors. The CI team, a sergeant and a specialist, had no choice but to detain the doctors and conduct an investigation. They interrogated them and a few days later cleared them, finding nothing that would indicate their being Taliban. Then the commanding General of KMTC fired Doctor A and rehired several of the Pashtun doctors.[94]

Rocks and Shoals

There was increased tension in the command group but I managed to stay away from the normal staff gossip because I wasn't an officer and I was busy with forty-three interpreters and their problems. My involvement with the ANA KMTC operations officer, who like our own officers seemed to believe that security was an intelligence function, rather than an operations function, which it is, and the demands of the doctrine section for more translations from the interpreters who wanted more money for more work, kept me oblivious to what was going on in our own unit.

Colonel Knight walked up to me one Friday after lunch and asked me if I would like to shoot a game of pool. The confluence of shooting and Colonel Knight had often flowed into my thoughts but pool was never part of the fantasy. Trying to be diplomatic I agreed to his strange request and we started walking towards the Morale Center near the PX. As we walked he began to talk about how the strain of combat was starting to affect the members of our unit. I listened with growing concern that he was on the verge of flipping out himself because we had not been in any combat that I knew about. He continued on that he was very concerned about the CO who in his estimation was about to fold up under the pressure of combat fatigue.

I was paying close attention to everything he was saying now. When we got to the Morale Center he got some pool balls from the desk and soon we were shooting 8-ball while he continued to talk about the impending train wreck with our chain of command. Because of his earlier stunts at the beginning of the deployment I didn't trust him one bit but this conversation was beginning to alarm me because it was slowly becoming apparent that he wanted me to get involved somehow for the sake of the unit and its mission.

After a couple of games, I won one and he won one, he said he had had enough and wanted to continue our conversation outside. I wanted to go to the PX and get some Gatorade and a few snacks but couldn't see any way of breaking contact with him so I went outside hoping that he would cut to the chase and tell me what he wanted from me.

We stood outside and he continued for a while longer talking about how we had to make sure that we had a strong chain of command that would ensure mission success and the safety of our troops. I agreed that this was our duty and was there anything else that I could do to support those goals. "Yes," he said, "you are the oldest combat veteran in our unit and you can see now that we have a serious problem. Maybe you should go to General M and tell him about our CO's impending breakdown so that he could help him by taking some of the pressure of command off him before he loses it."

As I stood there listening to Colonel Knight it finally dawned on me what was going on. As a sergeant I had seen a few attempts by an ambitious sergeant to get rid of a less motivated or incompetent sergeant so he could get the job but this was the first time I had been exposed to blatant 'officer' coup politics. This LTC, who I knew had lied to us about the situation at KMTC when we were getting ready to deploy and had caused me to waste hundreds of hours trying to prepare our unit for a situation that didn't exist, had been weaving me into his web for another foray into his own private world of ambitious fantasy. I was getting mad now but decided to keep it civil because he might be dangerous.

"Sir, I'd like you to remember that the UCMJ might consider this kind of talk dangerously close to mutiny in wartime. I think you should do what you can to help the CO bear the weight of his responsibilities so we will have mission success and all go home safely at the end of our tour." He gave me a withering look and walked off so I went to the PX and got the things I wanted.

When I returned to my tent I stretched out on my bunk, watching a movie on my portable DVD player and thinking about Colonel Knight's plot. I considered writing an MFR or reporting his conversation to the CO but we didn't get along very well in the best of times and I was sure that Colonel Knight would lie and deny everything. I could go to the CSM but he and I had a very cool relationship since he took my job back at Fort Carson, he for his own reasons and I because I thought he was clueless. I finally prayed and gave the situation to the Lord and fell asleep.

Or what king, going to engage another king in war, will not first sit down and deliberate whether he is able with ten thousand to meet the one coming against him with twenty thousand? Luke 14:31

Chapter Six: April 2004

Mail Calls

We got the word that Lee Ermey was coming to the KMTC range at Gharib Ghar. He was a former Marine Corps drill instructor who had gone to Hollywood and now he had a TV show. His show 'Mail Call' was filming a segment of him firing Russian weapons and we were invited to watch. Range control made him wear the IBAS which he didn't like.

After he finished shooting for the cameras he mingled with the troops. He made a few unprintable remarks about those who opposed our mission in Afghanistan. Then, he was asked to repeat a famous line from his most famous role as a DI in the movie, "Full Metal Jacket," which he did verbatim to the amusement of the troops. I couldn't resist a little thrust and parry and told him that he should stick around and we would make a soldier out of him. "No, I would be afraid of being corrupted," was his quick reply before turning his attention to some soldiers who wanted their picture taken with him.[95]

Marching Season

The morning staff meetings were starting to focus on a particular problem. Each one of our staff members had an Afghan counterpart who they spent much of each day with, working on the daily activities required for the running of a large army base while at the same time overseeing a massive construction effort. The problem was that the Afghan officers were no longer available after lunch each day because they were out marching on the parade field. ‘Marching Season’ was what our guys were calling the daily practice for the big parade at the end of April.

General A, who led from the front, was out there every day at the head of his staff section, marching up and down the parade ground that was next to the Five Story. The ANA style of dismounted drill was right out of the Russian manual. They slapped the soles of their feet on the ground with a straight leg stride that had to hurt.

General A was at least 55 years old and I could see he was hurting but he was determined that when KMTC marched in front of President Karzai and the rest of the Afghan political and military leadership, they would look like the Russian Army marching past the Kremlin on May Day. Around and around the parade field they went for several hours each day. After that ordeal there wasn't much energy for work. The CO had made several appeals to General A but the General marched on, knowing that their showing at the parade meant more for his reputation than all the great work we were doing daily at KMTC.

Force Protection

I was still having to deal with the security procedures established by the people who had come here before me. I'm sure that when they made the rules they made they seemed like good rules. I saw most of them as security driven by people who had fallen under the 'force protection' mentality that had taken over most of the military. Most of the soldiers in Afghanistan were restrained from taking any kind of offensive actions for the sake of 'force protection.' We isolated ourselves from the people in our HESCO Forts. We never went out at night when most insurgency operations take place and we were proud of our low casualty rates at the time, but that would change after we left.

The MPs wouldn't allow the interpreters to bring thumb drives into the Alamo. I considered all aspects of their concerns and then wrote a memo thru the Garrison Commander LTC Randy Davis, rescinding that policy for two of the interpreters, JS and JW. I would do the same for the others eventually but I wanted the MPs to take small bites of this new food so they wouldn't choke.[96]

Graduation

There was a Kandak graduation coming up and there was going to be a big show. I decided to take a break and video the event. Preparations had started several days before and on the morning of graduation day the area around the Five Story building looked good. About an hour before the ceremony the troops were marched onto the parade ground for some good old fashioned hurry up and wait. There the officers busied themselves with last minute adjustments while across the tarmac the guests began to arrive.

Several American generals with their personal security people arrived. They gathered with some of the colonels and chatted while their body guards fanned out and formed perimeters. These body guards wore helmets and body armor. They were all equipped with the latest add-ons to their M4 carbines and they all had stylish sunglasses. It was a little confusing for the rest of us, Afghan and TAG personnel, because we had to move through their perimeter to carry out our duties. As you approached one of the guys, you couldn't see his eyes which might give you a clue as to his intentions. Since they didn't stop us from walking through their perimeter eventually everyone ignored them and the area around the reviewing stand quickly filled with people.

The KMTC command group with our senior officers arrived and after some meet and greets with the visiting dignitaries they all moved to the covered area of the reviewing stand and the ceremony started. One of the ANA soldiers came to the microphone and prayed a long prayer in the manner of the Muslims. He sang with a nasal sound in what was probably Arabic. At the end all the Afghans lifted up their hands and then rubbed their faces and the prayer was done. After this General A gave a speech and when he was done it was time for the pass-in-review.

I enjoyed the way the Afghans had managed to resist our trying to make all things American; they marched like the Russian Army. The way they slapped their feet when they marched passed the reviewing stand made my feet hurt. They really looked good though and we were all proud as their performance somehow reflected on us and this would be all most of the generals would ever see of our collective effort.

When the parade was finished there was a pageant. Some ANA soldiers rappelled off the second floor of the Five Story building and started shooting at some others dressed like Taliban. They moved toward each other shooting, then dropped their weapons to engage in some hand to hand combat. The ANA vanquished their enemies and the crowd clapped their hands.

Awards were next and General A asked various officers to give an award to one of the Kandak soldiers being recognized. I was startled when I heard my name being called on the microphone. I looked up at the reviewing stand and General A was motioning me to come to him. After looking behind me to see if he was talking to someone else I trotted up to him and took a certificate which I gave to an ANA soldier who shook my hand. He then turned around and holding the certificate over his head shouted something in Dari before running back to his place in the formation.

At the end of the award ceremony came the last act, the 'man dance.' The Attan is the traditional dance of the Afghans. The dancing was done by volunteer soldiers from the Kandak. The camp barber kept them in step with the syncopated beat of his drum. He would shout, "Attan," at the soldiers when he felt they were slacking. Occasionally a few of our guys would join in and I gave it some serious thought but I could never figure out the beat or the moves.

Intel Ops

Steve contacted me by email. The Ishmael case was back on. His code words alerted me to a visit requiring Ishmael's presence in two days. I went to Major NS and asked him to activate the recall procedure. I decided to have the meeting with Steve and his people in my office so we could avoid exposing Ishmael to the whole population of KMTC. Major NS showed up with Ishmael and I escorted them through the MP checkpoint.

I was embarrassed that Major NS was searched along with Ishmael. The ANA were supposed to be our allies and we didn't search them when we went to their offices and they didn't search us. Searching them at the Alamo, a camp inside KMTC not only didn't make sense, it was one more insult to the people whose hearts we were trying to win.

Ishmael smiled and gave me a big hug. JW tensed up like he was preparing himself for Ishmael to explode, a suicide bomber. I didn't have any tea so I offered Major NS and Ishmael bottles of water and some MRE candy which they both accepted. We sat down together and they asked me about my health and the health and well-being of my family. I did the same and after we finished the meet and greet, Ishmael started to tell me about the latest happenings in his part of the world. I listened to the first offering and then asked JW to tell him to wait because we had some important visitors coming and I wanted them to hear everything.

As if by some miracle there was a knock on the door and in walked Steve, followed by an older bearded man. Steve introduced me to 'Tony' a colonel in the OOGA. I introduced them to Major NS, Ishmael, and JW. They sat down with us while I gave a brief summary of our activities with Ishmael, the CI team, and the OGA.

I told Tony that Ishmael had given me some information that was credible in itself and I was hoping that someone would listen to him. "That's why we're here," said Tony. I asked if he would like to talk to Ishmael alone. He nodded and so I took Major NS with me and we went over to the interpreter ready room after telling Steve where we would be.

After about an hour Steve came and got us. He and Tony were just finishing with Ishmael. They had worked out a contact plan and Ishmael had been given some simple tasks that would serve to establish his credibility. I had coached Ishmael not to ask for money right away. That and his answers to Tony's questions about his part of the world had established enough credibility for him to have an entrance back into the 'secret' world.

Steve and Tony left and JW and I escorted Major NS and Ishmael to the checkpoint. Ishmael gave me a big hug and I returned it. I had come to like him. He was playing a dangerous game and I respected his courage or craziness, hard to tell. He appeared to like me especially after I had done a personal favor for him. It had to do with the memory of his father and I saw myself having similar sentiments. That was the last time I saw Ishmael.

The Friday Follies

Friday was our day off. There was a weekly intel briefing every Friday morning up at Brigade and I attended it at the CO's directive. The briefing room at Brigade was just large enough for a U shaped table that had places for the General, his personal staff, the 45th Brigade staff, and the commanders of the various components of the Combined Coalition Joint Task Force Phoenix. My CO had a place at the big table and I, his intel guy sat directly behind him against the wall along with the other underlings, mostly eager young captains and majors who were part of the staffs of the subordinate commanders.

This was the brigade intel section's big day and they made the most of it. Major Bobby ran the briefing and whenever I saw him I averted my eyes so he wouldn't have to acknowledge me. He was still smarting from his defeat at the hands of his own General and I knew that he would try to get his revenge at the first opportunity. Usually I would try and antagonize someone like this but I guess I was growing up a little because I didn't want my personal feelings to interfere with our unit's mission.

After the long and drawn out explanation of their pattern analysis with Power Point slides of concentric circles and symbols that were supposed to graphically portray a clear pattern of enemy activity but instead showed that there was no pattern and therefore a waste of time and effort, the sergeant in charge of dismounted patrols around Camp Phoenix gave a detailed account of the week's patrols in the local villages. He showed photos of the homes they visited, the front door only, since they had enough sense not to try and go inside. They were numbering all the houses and keeping track of who lived in each one.

This was a good beginning to gaining a picture of the population around the camp. Afghanistan didn't have a province or national ID card so there was really no way of establishing identification like we have in the U.S. but it was better than nothing. What they didn't know about one of the villages and probably don't know to this day was a secret that I didn't find out until several years after I returned home.

I don't know if the General was buying this presentation as actual 'intelligence' or if he was just happy that his staff was staying busy. To me this was just another example of how certain soldiers with college educations managed to stay out of the actual fighting by casting their 'work' as 'classified.' They remained in the headquarters like the court jesters of old, good for entertainment but useless for anything resembling warfighting. Confusing effort with results had come full circle at Camp Phoenix where most of the soldiers and civilians were supporting each other with housekeeping duties. Those of us who actually had a mission were regarded as troublesome people, required to be there but an impediment to the important mission of running Camp Phoenix.

When the village tour report was finished we came to the morning entertainment. The brigade intel section then gave a briefing on some cultural aspect of life in Afghanistan. Several of them were quite interesting. The night shift LT, who was very helpful to me, was also in charge of providing the weekly briefing on some aspect of Afghan customs and culture. Two of my favorites were the Kuchi and Buzkashi.

Caravans and Rambo

The Kuchi were the nomadic people that could be seen passing through with caravans of camels and herds of sheep. We had several families of them camped in the field right outside the camp. They lived in tents and their women were the only women we saw that were not wearing the Chadori or 'Burka.' The Kuchi were the main Afghan characters in James Michener's book, "Caravans.[97]" It is a great read for someone who wants to know a little more about Afghanistan.

Buzkashi is the Afghan version of polo. When the movie about 'Rambo' in Afghanistan came out the west got a glimpse of this 'sport.' When you really take a good look at Buzkashi you have to wonder about the wisdom of going to war with a people who do this for 'fun.' Fortunately Buzkashi seemed to be more popular among the Northern Alliance people than the Pashtun but the Pashtun had their own diversions which were just as wild so it was better to be friends with all Afghans.

Both of these briefings helped me to understand a little more about the environment we were operating in and I wished that they were mandatory briefings for the rest of the brigade but several of the slides were classified 'secret' for some reason and so only the 'Friday Follies' invitees were allowed to see them. By now I was sure that the Brigade S2 shop was clue free about what was going on except for a series of incidents in Kabul that were suspicious but appeared to be random acts.

Conspiracy Theories

There had been a series of explosions in various parts of the city. They were caused by explosives concealed in pressure cookers, the Al Qaeda or Taliban style of bombs currently being used in Afghanistan. The last incident involved some agents of the National Directorate of Security (NDS) who were killed handling one of these bombs. Major

Bobby postulated that the series of bombs of this type were a stratagem of one of the factions of the government that was devised to keep ISAF in Kabul so that the private armies outside of Kabul could not be disarmed. This conspiracy theory may or may not have been an original assessment but it was the only one he had while I was there that made any sense.

The Contractors

Carl told me that there was a man at the camp next to KMTC who wanted to talk to me about intel. According to him, the camp was used for the training and housing of President Karzai's personal security detail. It was being run by 'contractors' who were working for one of the old time private military contract companies. I went to see him by walking from the Alamo to the entrance of the other camp, a short distance away. We passed the entrance to this camp every morning when we came to work and I always looked at the wrecked ZSU-23-4 that stood in the rubble near the entrance. There was a large concrete building that had taken a few direct hits during the bombing and the rubble contained parts of Russian artillery and the Zeus.

I was always fascinated with this particular anti-aircraft system because it looked so powerful. It had four 23mm automatic cannons on a rotating turret with a radar dish on top and it seemed like the airborne soldiers greatest threat on the battlefield. We had studied photographs of this weapon in the Long Range Surveillance Leaders Course and now I was getting to see it up close. It didn't seem to have had any effect on the planes that bombed it to pieces. Probably it served to announce a target to our precision bombing capability rather than a threat.

The guard at the gate didn't speak any English and I didn't bring an interpreter with me. He offered me his radio and I announced my purpose to someone speaking Australian English. The guard pointed the way and I walked up the road between the HESCO walls. I found the headquarters and knocked on the door.

I was greeted by the Australian who turned out to be from New Zealand. He stuck his head into one of the offices and a bearded man of about fifty years old came out and introduced himself as 'Mac.' He told me a little about what they were doing here which was to train President Karzai's bodyguards. He told me the President insisted that the men be from every ethnic group rather than just Pashtun like the President. 'Mac' was impressed with this and was obviously loyal to the man whose life he was helping to protect.

Carl had told me that 'Mac' had been in Delta Force. I was able to elicit a confirmation from him so I told him I was from Austin, TX where Colonel Charlie Beckwith, the founder of Delta, had retired. He enjoyed my stories about the colonel's visit to our unit and the Texas Drug Czar fiasco. He seemed touched that I had helped the colonel get a spiral staircase moved to his new house just before he died and said, "God bless, you."

After establishing our bona fides he told me that he would like to attend the intel briefings at Camp Phoenix. He had an expired 'Top Secret' clearance and he asked me if I would take a copy to the S2 and get him an invitation to the briefings. I said I would be happy to do it but would he like to hear my take on the 'intel' at Camp Phoenix? He nodded so I told him what I had seen and heard since we got here. "There is no intel at Camp Phoenix or any place else in Afghanistan," I said. "There might be some in some deep dark hole up at Bagram where everything comes in and nothing goes out, but I doubt it."

He didn't seem that surprised but while I told him that I would be happy to do what I could to get him anything he wanted, I also told him that in my opinion, he would be wasting his time. I left and went back to the Alamo. When we got back to Phoenix, I approached one of the S2 NCOs and asked him to help me get permission for 'Mac' to attend the briefings and the next day he gave me a thumbs up. I went back and told 'Mac' it was a go. He attended one briefing and after that I never saw him again at Phoenix.

I continued my daily routine of gathering the latest 'intel' from brigade and boiling it down to a one page summary that started with the weather. After that I would put two or three incident reports from either Camp Phoenix or Pol-i-Charki. These were usually suspected surveillance operations. Then came the hot tips from around the city: car bombs, truck bombs, suicide bombers and assassins. The last section was incidents around Afghanistan.

One morning the big deal was a story about how saffron, a spice, was going to be the substitute crop for opium. Afghanistan was believed to be producing one half of all the heroin being used in Europe and the UK. If the government tried to stop the industry in places like

Kandahar and Helmand Provinces in the south, the people in that area would support the Taliban even more than they did now. This was just one more example of wishful thinking that was being passed off as intel. Before the day was over there would be people discussing the wisdom of the saffron instead of opium plan.

R+R

The word came down that we could have a three day trip to Qatar for R+R. The trip actually took a week with the transportation glitches. The troops could go to malls, visit the desert, and eat some different kinds of food, oh boy!

When I was in Vietnam I had a memorable trip to Hong Kong for R+R. Due to a major disorientation caused by lack of sleep and an overdose of alcohol I missed my flight back and spent a day under house arrest, no alcohol and no visitors, in a hotel there waiting for a flight the next day. The marine who shared my room and had a similar problem was astounded when I lowered some sheets that were tied together so that a local supporter of our efforts in Southeast Asia could attach several six packs of the local beer, for which I paid him a finder's fee. After reminding the marine that the worst that could happen to us was getting our heads shaved and being sent to the infantry in Vietnam, which is where we were going anyway, he saw my logic and shared several beers with me before we arranged with the desk clerk to break rule number two.

I was a new man now and those pleasures no longer had a hold on me so I decided to pass on the three day trip to Arabia. I was looking for at least one distinction on our deployment and this could be it. I would stay in Afghanistan longer than everyone else in TAG by at least three days. As it turned out I would be staying a lot longer than that.

The Big Parade

Toward the end of the month I began to track daily threats against the Independence Day Parade in downtown Kabul. I heard that several of our officers were going so I asked to go when it was being discussed at the morning staff meeting. LTC Kevin asked me if I had a death wish. I got the OK to go and made sure that that my batteries for my video camera were charged and that I had plenty of tape.

On the morning of the parade I joined the other parade goers and we drove into Kabul. There was lots of traffic, as usual. One of the interpreters in the lead vehicle directed us to the backside of the reviewing stand about a block away. We were directed to park in an open field guarded by Afghan police. We walked toward the reviewing stand. There was a line of VIP vehicles unloading near an entrance and there were a number of bearded westerners in the de rigueur costume of the time, Royal Robbins 511 khaki colored pants, shirts, and vests with an Afghan scarf around the neck. Wiley X sunglasses hid their eyes but it was easy to see that some of these guys had been around for a long time and had jumped on the gravy train when it came to town after 911. They all had M4 carbines on special slings and pistols in drop-rig holsters on their thighs.

Anyone with an SF background was certain to be hired and I saw quite a few men that looked like they had been in Vietnam before I was. A contract security guy could make more than $200,000 a year, tax free. This gold rush lasted for more than ten years. The congress was willing to spend more money on contractors than it was to hire more troops. The soldiers were also getting paid a lot more in Afghanistan than we did in the states. Almost everyone was benefiting financially from this war, except the unconnected Afghans.

All our guys that came along to be security for the rest of us were trying their best to look like these guys. They had the sunglasses, M4s, and pistols on their thighs. The CO wouldn't let them wear scarves. I'm sure that they were keeping an eye on the people around us but there were so many Afghans, in and out of uniform, that I gave up trying to figure out who was going to try and kill us and trusted that the Lord had our back or I wouldn't be here.

The main section for Afghan VIPs was about fifty feet to our right. There was a low wall between our section and theirs. Our section was full of generals and colonels from the various western members of ISAF and the U.S. military. The Afghan section included President Karzai and his Vice President Marshal Mohammed Fahim, the leader of the Northern Alliance after Ahmad Shah Massoud was assassinated by Osama bin Laden two days before 911. I saw Ismael Khan, the famous mujahideen commander from Herat gesturing wildly to some old friends, signaling for them to come and join him.

The parade started with a long prayer by a mullah. We all kept quiet out of respect for the fact that we were sitting in a huge crowd of Muslims who were known to become severely agitated when there was any sign of disrespect toward their religion. I scanned the area across the parade ground where all the military units were assembled in front of the huge mosque. There were security forces on the roof of the mosque and along the perimeter of the parade ground. I took video using the small screen so that it wasn't obvious.

After the prayer I noticed a Russian military convertible with a hand hold driving down the parade ground and coming to a stop in front of the reviewing stand; in it stood President Karzai, wearing his Karakul and colorful cape. He had devised his costume as part of his effort to unite Afghanistan as one people rather than the collection of ethnic groups that had been at each other's throats for as long as anyone could remember. He held his hand up to his eye brow in a salute and held it there.

Then from the other end of the parade ground came another convertible with another figure, this one in a dress uniform with a high peaked hat. It was Fahim, the Vice President and Minister of Defense. He faced President Karzai and also held a salute. Behind them were giant pictures of Karzai and Ahmad Shah Massoud, the 'Lion of the Panjshir.'

This was an Afghan pageant designed to explain the relationships between all the members of the fragile coalition known as the Afghan Transitional Authority. Fahim was definitely a power behind the throne, the leader of the 'Panjshir Valley Boys' a Tajik coalition from Northern Afghanistan. There were rumors around that Fahim had tortured Karzai during the civil war that followed the Jihad and that Karzai had escaped after a rocket from Hekmatyar landed on the place where Karzai was under arrest. Whatever their differences were in the past, they were putting a nice face on it and pressing forward while watching their backs.

The parade started and the first groups of marchers were men in traditional Afghan clothing carrying AK-47s, RPGs, and PKMs. There were seven groups; one after another they walked in front of the reviewing stand. They walked, not marched, like the guerrillas they represented. These were role players representing the seven major mujahideen groups that fought the Russians and the Afghan communists. The crowds in the stands and those lining the parade ground began to clap and cheer as they walked by. All the Afghan army generals in the stands were on their feet and saluting. This was one of the many ironies that day as most of the generals were wearing Russian decorations on their uniforms.

After the mujahideen groups came the ANA; across the parade field, unit after unit was called to attention and given the order to march. In front of the reviewing stand on the other side of the parade ground stood a number of drummers who would provide the beat for the soldiers to use to keep in step. I was trying to video everything with my small Canon camera. I had three batteries but they were small. My larger Canon had stopped working about a week before and I had sent it back to the states for repair. My smaller camera didn't have much zoom and I wanted to get some shots of General A and the KMTC staff when they marched by. The CO was particularly interested in giving the General some photographs and video of the event as a kind of 'hearts and minds' gift. General A and his staff had labored long and hard to prepare for this event and we wanted them to see the fruits of their labor.

Unit after unit marched by and finally it was the turn of KMTC. There was General A at the head with his sword and the staff on line behind him with a Kandak that was ready to graduate from basic training behind them. The crowd was clapping with us. This was our moment to share in the pride of the ANA whom we were committed to help.

Then came the armor; the Afghans had lots of Russian tanks, mainly T-55s. They also had lots of BMPs. There were enough running that they put together a tank company and a BMP company. We all held our breath as they slowly moved past the reviewing stand. There was a lot of smoke and noise and in the middle of it there was more noise as a number of Russian helicopters flew overhead, mostly MI-17 'Hips' and MI-24 'Hinds.' The Hinds were the most impressive and the target of Congressman Charlie Wilson during the Jihad. Once again we held our breath as they flew over, hoping that the maintenance crews had done a good job.

The parade finally ended and after the VIPs were hustled out of there we left by way of the back door of the reviewing stand and walked back to our vehicles. We checked underneath our vehicles for bombs and then joined the traffic jam headed for our part of town. The armored units were ahead of us and some of them had reached the end of their mechanical endurance. They were stalled on the side of the road or on the road itself, blocking traffic.

It took about an hour to get through the traffic and we got back to Camp Phoenix before dark. Rambo greeted us with his usual salute, his sons watching their dad at work from the doorway of his home at the gate to Camp Phoenix. He made it his personal business to deal with any threat to the Camp at the main entrance. A few years after I left he jerked a suicide bomber out of his vehicle before he could detonate the bomb in the back, a brave act by anyone's definition. I had a few encounters with him while I was stationed there and he remains one of my heroes of that war.

The Estimate

I finally finished the Intelligence Estimate for KMTC.[98] This was my magnum opus. I had followed a format in one of the intel manuals and it was now ready. I gave it to JW and he translated it into Dari. We did it together so he understood all my terminology which I tried to keep as simple as possible. In it I did a kind of IPB, intelligence preparation of the battlefield, which is a complicated way of describing all the possible ways and means for the enemy to attack you. I described the terrain and what weapons might be used and from where. I also covered the internal threats and possible vulnerabilities. When we were done I took it to Major NS and gave him a copy. I encouraged him to read it and maybe he did but the next thing he did was to give it to the operations officer who circulated it to the senior staff.

I came to work the next morning and there was a message for me to come to Major NS's office as soon as I could. I figured he was getting lots of compliments on 'our' Intel estimate so I hurried with JW to his office to enjoy his moment of triumph. When we got there the look on his face was hardly one of celebration. Sheer panic would be a better description. He began to sputter and stammer in Dari. JW had to ask him several times to slow down or repeat what he was trying to tell me. The gist of it was a line in the internal vulnerabilities section that stated that the leadership at KMTC was at least 90% Tajik and since the demographics of Afghanistan indicated that at least 40%, the largest percentage, were Pashtun, this might become a source of trouble in the future.

This did not go over well with the senior leadership and they were calling for the head of Major NS. By now he was begging me to tell them that he was not the source of that information and that he had nothing to do with its inclusion in the estimate. I was pretty sure we were in deep shit and told him that I would do what he asked me to do immediately. We went to the operations colonel and asked for some of his time.

He invited us in and I asked JW to tell him I was very sorry for causing any trouble with my estimate. He nodded and told me that the estimate was being studied by General A and we would have to wait and see what he said. He didn't offer us any tea and indicated we could go. We went back to Major NS's office but he was in a depressed state and didn't want to talk so we left him and returned to the Alamo. I wrote a memorandum for record, detailing the events and worked on other things until it was time to go back to Phoenix.

The next day I got a message from Major NS requesting that I meet him at the operations office. When JW and I got there we found several of the senior colonels sitting at a table. Colonel Z told us to sit down and they began by asking who wrote the estimate. I told them that I did. Then they wanted to know where I got the demographical information regarding the disproportionate representation of the Tajiks in the KMTC command structure. I told them that that was my own estimate. Then they wanted to know if any Afghan had given me this information and I told them no. Once again I claimed full responsibility for the document and even threw in that I told Major NS to distribute it before he had finished reading it because it regarded the security of the camp. This seemed to mollify them a little but they told us that the estimate was in the hands of General A and we would have to wait on his ruling. The last thing they said before dismissing us was that we were to return the next day.

I was expecting the CO to call me about the flap or bring up something in the morning meeting but all was quiet so I kept it to myself. I took JW and we went back to the Five Story after the staff meeting. The same group of colonel's was there along with Major NS. The mood seemed to be lighter so I sat down to find out our fate. The orderly served us all tea, a good sign. Then the operations officer began to speak. He went on quite long before he stopped so that JW could translate what he said. JW told me, "The General told the colonels, don't worry about it." The colonel got a little miffed because JW didn't translate more but JW wasn't intimidated and told me again, "don't worry about it." Major NS breathed a sigh of relief and everyone sipped their tea and waited to hear what I would say. I told JW to ask them if I could have the copies of the estimate, some salt, and a large glass of water. He translated my request while I watched their faces take on a puzzled look. They asked him why I wanted the water and salt, thinking perhaps that I was going to show them a secret way to make the estimate disappear. I told them I wanted the salt to make the meal a little tastier because I was going to eat the document. This got a big laugh and we were all friendly again.

The baptism of John, was it from heaven or from men? Answer Me.
Mark 11:30

Chapter Seven: May 2004

Statistical Intelligence

Major B the logistics officer had an expense fund of $5000 to be used for immediate necessities at KMTC. He and the other logistics people along with various straphangers looking for a trip into town would drive into Kabul almost daily to take care of the little things that had not yet entered into the logistical pipeline that was overseen by 45th Brigade and OMC-Alpha. Their interpreters would direct them to places like 'Chicken Street' and 'Paper Street' in downtown Kabul where the interpreters would bargain with the vendors and probably get a middleman fee in the exchange.

Of course more trips into town meant more risk and the CO was looking for some way to limit them, believing that the more trips they made, the more likely it was that they would be attacked. Actually this was a problem because many of the troops who came to Afghanistan during the war expected to see Afghanistan and when they found themselves restricted to the base they did what American soldiers have done in every war and found ways to disregard those restrictions and make 'mission essential' trips to any interesting place nearby.

The CO had become quite enamored with the statistical intelligence slides that the 45th S2 was presenting as 'intel' at the nightly BUB and the weekly intel extravaganza. The PowerPoint rangers on the S2 staff had been charting all the incidents in Kabul since the 45th arrived and displaying their work each week with any new events added to their slides.

One day he directed me to do a statistical study of the same events on which the 45th was basing their predictive intelligence products. This forced me to actually look at a copy of their slides that the S2 LT on the night shift graciously allowed me to download to my thumb drive. After an intense study of the data I determined that there was no discernible pattern of anything more than sunrise and sunset.

What the colonel really wanted was to rein-in the many trips to town by the bored members of his command. He wanted me to log in every trip to town over the last several months and chart the pattern of trips. Then he wanted me to overlay this product with the enemy activity events so he could show the logistics guys that it really wasn't safe for them to go to town, ever.

This project did little to inspire my enthusiasm as I knew this was all a command and control scheme. When I reluctantly began gathering the data I found that our trip to town records were both inaccurate and incomplete. If I reported the bad data the colonel would have a field day with the ones responsible for tracking the vehicle activity so I took what they had and tried to figure how to do this dance. Every week the colonel would ask me how I was doing on the project and I would tell him what he wanted to hear, "I'm on it sir, it's almost done." Meanwhile I was unloading on Major Shawn, the Ops officer, and he had one of the Ops sergeants help me with the graphics on the PowerPoint slides. This project would continue to annoy me because it was so contrived and it would annoy the colonel because it was never quite done.

Trouble in the ANA

Major NS needed help. He was quite distraught and began to pour out to JW whatever was bothering him. I waited for him to stop so I could get a translation but he kept talking in a very agitated manner for at least five minutes. Finally he stopped and JW told me the problem. Major NS had bought a small house near KMTC in a military housing project. He moved his wife and children in and was now doing like the rest of us do, making some home improvements. He wanted to put a toilet in and was building a room for it. The day before while his son was working on the project, some ANA MPs forced their way into his house, beat his son and told them that they had to have a permit from COL R, one of the ANA colonels at KMTC. The permit was an expensive one but what Major NS was most angry about was the way these MPs had violated the sanctity of his home and for this, his honor.

His first remedy, and he needed my help for this, was to get an AK-47 from supply and kill the officer who had engineered this home invasion which was really an extortion scheme. I sympathized with him, having considered the shooting of a few senior officers to be the best solution for a multitude of problems. Fortunately I had been brought into captivity by my Savior so I had to reject these thoughts and look for a better solution to Major NS's problem. I looked at him while he ranted to JW. He was trapped in an impossible situation—Afghanistan. I wanted to tell him about my Savior and what he had done for me but that would have to wait for a calmer time.

I took out my notebook and he gave me a sad smile as I showed him that I was writing down all the details as JW gave them to me. I wrote notes about the incident, the: who, what, where, when, and why. When JW was finished and I had finished writing I told Major NS that I would take this problem to the CO. He got frightened and begged me not to tell him who it was that complained. He explained that there were a lot of others who had been shaken down by COL R. If General A were to tell him to stop without knowing who complained, COL R would have to do so or he would have to retaliate against everyone in that housing tract. That would start a small war.

I wrote a memo to the CO about the situation and sent it to him by email.[99] This was the way we did things now in the army. Instead of having face to face conversations we 'shot an email.' However, I did accost him in the courtyard that afternoon as he was getting ready to leave and told him that my counterpart was in fear of his life, asking for his discretion in the matter with General A. He indicated that he understood and would take care of it. I guess he did because two days later Major NS greeted me with a big smile and told me things were better on the home front.

Connecting on the Internet

I was talking with my wife on the internet each morning for about fifteen minutes. I would give her my prayer requests and she would assure me that all was well on the home front. I asked JW if he would like to use my phone program to talk to his mother in Pakistan. He was very happy and talked to his family at least once a week. I ran into one of the interpreters in the courtyard, we'll call him Khan. He was returning from an assignment and I asked him if he wanted to come to my office and have a snack. He did and followed me. We ate some MRE food and talked. I asked him about his family. He told me that his father had abandoned his family in Pakistan and moved to the United States. His mother and sister still lived in Pakistan. I asked him if he had ever talked to his father on the telephone. He said he had a number but he had never used it. It was early in the morning so

I asked if he would like to call him on my internet phone and he said yes.

He gave me the number and I dialed it on my laptop and gave him the headset. I watched his face change when someone answered. He began speaking in Pashto. Soon there were tears pouring down his face as he and his dad reconnected after years of separation. They talked for about fifteen minutes and then he took the headset off and said they were done. He thanked me several times then left.

Baptism

A few days later Khan asked to speak with me in private. I invited him into my office and closed the door. "Sergeant Murry, when I was in Pakistan in a refugee camp, a European worker told me about Jesus. After a while I told him that I believed that Jesus was the Son of God and prayed to Jesus. Now I want to get baptized. Can you baptize me?"

Wow! A lot of thoughts flooded my brain. Where could we find water? How could we do this without being seen? Could we even do it here at KMTC? Could we do it at Phoenix? I told him that I would be happy to baptize him but it would take a little while to figure out how. When I got back to Camp Phoenix I talked to the chaplain about it in general terms and he shied away from any speculation on my part that he could be involved in anything like that.

He had already asked me to leave his Bible study when I suggested that we stop studying James and take up Galatians. Like most good Baptists, he was more interested in exhortations to keep the law. Luther wanted to burn the book of James. I wasn't that bold but I did mention the verse that Martin Luther used to turn the world upside down, Galatians 2:16, "And knowing that a man is not justified by works of law, but through faith in Jesus Christ…" Luther was kicked out of the Catholic Church and I was kicked out of the Bible study.

General Order Number One

I continued to look for the way. One of the KBR maintenance workers I had gotten to know through MSG Bill was a Christian. I asked him if he could help me find enough water to baptize a small Afghan. He told me that he would help me but I could see that he was pretty nervous. When I asked him what was bothering him, he told me that he would be fired if we were caught baptizing an Afghan. "General Order Number One," he said.

He could get fired, I could get court-martialed and Khan could get killed. I knew what he was talking about. GO1 said no proselytizing but this was a request from a believer, not an attempt to convert a Muslim. I knew that we had to keep this a covert operation but I didn't think I was in violation of the order. The KBR worker agreed to help me. All we needed was enough water to immerse a small man for a moment. He told me to check with him in a few days and I went back to my tent confident that this was well within my capacity as a low visibility operator.

The Norwegian Hit

The camp alarm went off at about 2000 that night. The loudspeaker announced that this was a real alert. The guy who slept next to me said that he had to go out to the vehicle parking area and get his helmet. He was one of many young, and old, soldiers who had grown complacent and began leaving their equipment in their vehicles. I told him to stay where he was and gave him my helmet because I didn't think it would be safe for him to be walking around in the dark with a bunch of people who hadn't been under fire before. Friendly fire isn't friendly and there was a real possibility of it in this camp. After about thirty minutes of waiting for the mortar rounds to come in I decided to go to sleep and did.

The next morning at 0400 I went down to the Intel section at Brigade Headquarters and LT Fuzz briefed me on the events surrounding the alert. A Norwegian patrol from ISAF was ambushed on the Jalalabad Road between KMTC and Camp Phoenix. One man was killed by an RPG. That was all the information they had at that time so I left and went back to my tent, prepared my morning briefing paper and went to breakfast.

After breakfast we assembled at the car park where we were informed that no one was going anywhere. We were locked down until further notice. The hit on the Norwegians had unnerved the leadership at 45th Brigade or higher and they decided that it was too dangerous for us to leave the camp. I saw the CO nearby and complained that this was going to greatly embarrass us with our ANA counterparts.

"They're going to think we're cowards if we don't go to work and besides I came here to fight, not to hide at Fort Fumble until its safe." He gave me one of his mean looks and said we would go back to work when the 45th said so. The next day they lifted the travel ban and we went back to work.

As I rode to KMTC that next morning I looked for signs of the ambush but saw nothing unusual on the road. I briefed what I knew about the incident at the staff meeting. The Colonel used the incident to remind everyone that we were in a war zone and that there were still bad guys around. I left the Alamo and went to the interpreter's building and found JW.

The Great Game II

We went to Major NS's office and began our work. Instead of the usual requests for equipment and people, Major NS asked me if I had heard about the ambush on Jalalabad road the day before. I told him that I had heard a few things and I asked him what he knew. He told me that he had a source in a village near where the ambush had taken place who knew who did it. I was a little taken aback because of what we had gone through with Ishmael, but I asked him to bring him in for a talk. He said he would and told me to meet him after lunch.

I went back to my office and did a search on the internet for any news stories about the incident. There were a few and I took some notes. After lunch I got JW and we went back to Major NS's office. When I walked in there was a man sitting with the Major in traditional Afghan clothing, a turban, long shirt, pants, and sandals. We greeted each other and sat down. Major NS introduced me to a man we'll call Badek, the man who said he had information about the ambush.

Badek had been in the 'Game" for a while. I asked him to tell us about his background. He told us that he had been an intel operative with two different mujahideen organizations. This guy was smooth because many in them were enemies to the death; Badek had worked for both of them and was still alive. After the U.S./Northern Alliance arrival in Kabul, he kept his hand in the game for a while but was 'burned' in some incident and now he was 'freelancing.'

I asked him to tell us about the ambush. He told us that four or five men left a nearby village, walked west along the Kabul River to some drain ditches that led to the Fanta/Coca Cola bottling plant which was on the south side of the Jalalabad Road. They hid behind some piles of dirt and when the Norwegian patrol passed by they fired three or four RPG rockets at the convoy and escaped the way they came. He said they had all gone to Pakistan to lay low for a while.

He gave us a great deal of information about past events involving the anti-coalition militant activities around Kabul during the past two years. Once again, the information seemed credible but I had no way of gauging his reliability. My relationship with the CI team had deteriorated to the level of contempt on my part so I wrote a report and emailed Steve with a sanitized version of this information.[100] He told me he would be at KMTC the following day to talk to Badek. Steve showed up with a CPT G and G interviewed Badek. They talked to him for several hours and while they were talking, Steve asked me what I hoped to get out of this. That surprised me.

I thought quickly however and told him I hoped to get a pepperoni pizza from the ISAF base across the street. He laughed and said he would be right back. I got a radio call to go to the Five Story and when I returned, there were two boxes of pizza sitting on my desk; thanks, Steve, they were delicious. I did offer them around the office but nobody took me up on a slice of pizza. I thought that was strange until I found out that most of the TAG group was finding reasons to go to ISAF almost every day and they were sick of pizza.

Sometime later I found out about what happened with the information. Captain G presented his report to the ISAF intel people and they considered it a good possibility but eventually they arrested someone else and considered the case closed.

Train Wreck

The interpreter who asked me to baptize him was asking me every few days about when we were going to put him in the water. Finally the KBR worker who had agreed to help me said that everything was ready. I went with him to the maintenance area at Camp Phoenix and he showed me a large barrel full of water. It was perfect. I told him that we would do it the next night, right after dark.

The next day I found him and asked him if he could meet me at Camp Phoenix that evening to be baptized. He smiled and asked what time. We worked out the details and I went about my business. I visited Major NS in his new office which was in the same building as the interpreters and we had a pleasant visit. I ate lunch with JS and we discussed the computer training program that I had gotten from CD. One of the interpreters, a doctor by education, had studied the database program and had constructed a pretty sophisticated database for JS to use in managing the interpreters; that made me feel good.

I went back to my office and did a little work. The man that Khan interpreted for stuck his head in and said hi. I knew he was a Christian so I invited him to help me baptize Khan that evening. I told him the story of how he asked me to baptize him and he seemed fine with it. He left and I continued to work. About ten minutes later I got a call on the radio telling me to go the CO's office. Figuring it had to do with some memo I had sent him I went upstairs where I was met by the CSM who told me to follow him to the CO's office.

He was sitting at his desk with a tight smile on his face. I had only seen him smile once or twice before so I thought that something must be up. "Did you plan to baptize one of your interpreters tonight at Camp Phoenix?" he asked, still smiling. Bottom line up front question deserved a similar answer. "Yes sir." He then smiled at the CSM and said, "you are to gather all your personal belongings and take them with you to Camp Phoenix tonight, you are not coming back;" I was devastated but I managed to say, "yes, sir, "before asking if there was anything else. He dismissed me, still with that tight smile on his face, telling me to report to his quarters after dinner that evening.

I was in a bit of a daze as I walked down stairs. I saw the man I invited to help. He looked stricken. "I'm sorry," he said. "I told the CSM, not thinking that he would run in and tell the colonel. I'm sorry," he said several times. I told him that I had been relieved and that I was going over to the interpreter office to get some things. I wanted to tell JS that I was leaving and that I would try to stay in touch when I found out what I would be doing next.

The colonel had assigned a sergeant from operations to accompany me. He and I were on good terms and he looked the other way when I said goodbye to JS and all the interpreters who were there. I gave the TV and VCR to JS and wrote a statement so stating and signed it. I told him that I would let him know where I was so that I could give him and the rest of the interpreters' letters of commendation for the period of time they worked for me. I hugged him, my eyes tearing as I left what had become the most important thing in my life for the past six months, the Afghan interpreters who had become my friends. I told JW to work with whoever they sent to replace me and that we would stay in touch.

I rode back to Camp Phoenix in a daze. I went to my bunk and lay down. The CSM came in and told me to report to the colonel's quarters after dinner. He and the CSM were sharing a building near my tent. I went over and the CSM told me to sit down. The CO went off quietly about my trying to baptize an Afghan and how if the word got out we could be attacked. That was certainly a possibility so I held my tongue. The CO then told me that he was relieving me of my duties with TAG and that I would be reassigned to the 45th Brigade.

Combat Stress

Then he said, "I'm sending you to Bagram Air Base to be evaluated by the Combat Stress Unit." That brought an involuntary smile to my face. Colonel Knight had tried to manipulate me into telling the General that the CO was suffering from combat stress and now the CO was telling me that he thought I was suffering from the same thing.

Oh the irony! I decided to keep that to myself since Colonel Knight had been relieved by the CO a month prior for some close held offense and I couldn't see any benefit in bringing that up at this point in time. The troops were all getting a little restless after six months of an unusual assignment and I didn't think that fighting with the colonel about this would help the overall situation. The colonel stopped talking so I asked him if that was all. He turned to the CSM and asked him if he had anything to add. The CSM asked me if I had anything else to say but I ignored him, saluted the CO, and left. One of the sergeants from RI had heard about what had happened and he told me to fight this as an attack on my religious rights but I felt it would be better to take whatever they gave me and support the chain of command.

Bagram Air Base

The next day, as directed, I reported to sick call and told them I was being sent to Bagram to be evaluated by the Combat Stress Unit. They processed my paperwork and told me that we would be leaving in a convoy at noon. I had a book to read so I made myself comfortable until it was time to go. I caught the medics checking me out several times out of the corner of my eye but ignored them. The call came to load up so I found a seat in one of the pickups and we drove out of the camp and headed for Bagram.

The road from Camp Phoenix to Bagram had a lot of local traffic and I took some pictures to keep busy. We were being escorted by a couple of HUMVEEs; one had a .50 caliber machine gun and the other had a Mark 19 grenade launcher so I figured we were a hard target. IEDs hadn't become popular in Afghanistan at that time so I felt pretty safe.

We drove past the OGA 'secret' site and the traffic by then had thinned out. We were now on the Shomali Plain. There were Kuchi camps and camel trains out in the desert on either side of the road. Every so often we would pass a destroyed T-55 or BRDM near the side of the road, testimony to the fierce fighting when the Northern Alliance, the Taliban, the mujahideen, the Afghan government and the Russians all fought for this approach to Kabul.

The trip to Bagram took a couple of hours. I had been here once before, riding shotgun on a convoy, so I was vaguely familiar with some of the landmarks. As we got closer I could see air activity as A-10s made their approach, returning from an air support mission somewhere down south. The MPs waved us through the checkpoint and we drove down the main street toward the American medical facilities. I could see the old control tower that ODA-555 had used as an observation post when they assisted the Northern Alliance troops in taking Bagram back in 2001.

The Combat Stress Unit

We turned off the main road into the medical compound and dismounted. The medical NCOIC pointed at a building nearby and told me that was where I wanted to go. I walked across the dusty traffic area and entered the building, a SEA hut with a reception area inside the door. I handed my medical records to an Air Force sergeant and told her that I was here to be evaluated for combat stress. I smiled again when I said 'combat' because it seemed so ludicrous given that there wasn't any combat where we were. She told me to have a seat and took my records through a door behind her. There were a couple of soldiers from the 10th Mountain Division who I guessed were there for evaluation also. I knew that some of the combat units had seen some fighting down south and along the border. I didn't say anything to them and they ignored me.

I had never heard of a combat stress unit before. We didn't have such things in Vietnam. This was all part of the modern military at war. In the years that followed, especially in Iraq, these kind of psychological treatment facilities would become busier and busier as soldiers began to stress out after multiple deployments. The modern military medical doctor's answer to the demands of the service to keep up the fighting strength, the numbers, was to prescribe medication, anti-depressant and anti-psychotic, to the soldiers. The unintended consequences of this practice are beginning to manifest themselves as the war winds down. Whenever one of the possible side effects of an anti-depressant medication includes suicidal thoughts you have to wonder if the doctor prescribing it was humming the theme song from 'MASH' as he filled out the form.

I heard my name called from the back of the building so I rose and walked through the door the sergeant was pointing at. I continued down a hall until I passed an open door. A voice told me to come in and when I did I saw an Air Force Lt Col sitting at a desk staring at me. There was a chair in front of his desk so I sat down and waited.

The AF officer seemed a little nervous as he read through my medical records. Finally he looked up and asked me if my combat stress had caused me become dangerous. I thought that was a strange way to start a psychological evaluation but since he asked me a direct question I decided to let him decide by saying, "Do you think they would let me carry this pistol," pointing down to the M9 in the holster on my right side, "if I was dangerous?" That got his attention and he looked startled to find out that I had a weapon. His hands began to shake and it appeared to me that he might need some relief for his own combat stress, so I told him, "Look, doc, I'm OK. This is a crazy situation and if you give me a minute I'll tell you why I was sent here." I guess that seemed a rational enough response and he stopped looking like he was ready to bolt and relaxed a little.

I told him the quick version of what had happened while he took notes. I said that I was mystified as to why the colonel sent me here but it had been a strange war so far. When I finished he asked me a few questions regarding my current mental state; was I thinking about killing someone, did I want to hurt someone, was everything all right at home, etc. My answers seemed to be what he wanted to hear and after a few minutes, he looked up at me and said I was ok. He signed the form he had been writing on, handed it to me and told me I could return to duty.[101]

The Big PX

I picked up the rest of my medical records at the front desk, signed out, and left. The convoy was parked nearby and I found the NCOIC who said we had some time to kill and that I could go to the PX nearby. I took off up the road, hoping to get a hamburger at the Burger King that was rumored to be operating near the PX. There was such a building but it wasn't opened yet so I joined the line going into the PX.

Standing in that line were military people from all the services and coalition forces from all over the world. There were a few grunts from the airborne battalion stationed in Alaska that had been pulling security at some of the FOBs (Forward Operating Bases) around the country. They were, of course, mildly contemptuous of the rest of us, because they were the fighters, the reason we were there as far as they were concerned. I knew how they felt; but just as it was in Vietnam, the REMFs (Rear Echelon Military Forces) had no shame and were contemptuous of us being too stupid to get out of the fighting like they had, although they were smart enough not to mention it in our presence. Now they were called Fobbits, and there were many more of them, all doing essential war work inside the HESCO fort.

The PX was a trip. There were all kinds of diet supplements for body building, a major activity at this combat base. A large DVD section offered the latest movies at stateside prices but you could get the same movies at the bazaar for a dollar if you didn't mind someone in the theater shaking the camera or an occasional silhouette passing across the screen as someone walked in front of the camera on the way to the restroom.

There were plenty of consumer electronics to include TVs, portable music players, and cameras. The clothing section featured cold weather items and lots of PT gear. There were plenty of snacks and even a small book section. I bought a military history book for something to read when I got back to Camp Phoenix because it looked like I was going to have a lot of time on my hands.

A Massage Parlor—Really?

When I finished wandering around the PX I walked back towards the hospital. I passed a small strip mall. There was a shop that was selling sports equipment and another shop was advertising massages. That brought me up short. I couldn't believe that they were going to allow that kind of stuff in Afghanistan, didn't they remember anything? I peeked in and saw a young lady in civilian clothes working behind the counter.

There were a couple of infantrymen sitting on a bench outside so I asked them what this place was all about. They laughed and said it was a message parlor run by girls from one of the Russian Republics north of Afghanistan. "Go on in and get a massage sarge, it's a pretty good deal," one of them said. "My wife wouldn't like that," I said with a smile. "What happens in the Stan…" I heard one of them say as I walked away.

When I got back to the hospital, the medics were done with their business and we mounted up to leave. Of course they needed to go to the PX so we went back to where I had just been and spent another hour doing the Bagram shuffle. I sat outside most of the time watching the parade. The contractors in their 511s and sunglasses, some kind of special ops or intel types. Special Ops people in uniform now and shaved, strutted around like they owned the place. The majority of the people were the housekeepers, people stationed at Bagram taking care of people stationed at Bagram.

Finally we left and had an uneventful trip back to Camp Phoenix. As we rode along I started getting some pretty hateful thoughts about the CO. I considered the way he had treated me since that day that he had me argue with Colonel Knight about the combat patrols with the ANA that Colonel Knight had dreamed up as a way to alleviate his mid-life crisis. I was being bombarded with negative thoughts and at a certain point I began to call on the name of the Lord Jesus and pray that the CO would be saved. I had a big wound in my soul and the only way for it to heal was to pray. When we got back to Phoenix, I checked in with the CO and told him I had been cleared by the Combat Stress Unit. How he took that I couldn't tell but he did tell me to report to the 45th Brigade's Chief of Staff in the morning.

Tajbeg (Queen's) Palace-Darulaman
(authors photo)

Old Russian Officers Club-Darulaman
(author's photo)

And they said, Cornelius, a centurion, a righteous man and one who fears God and is well attested to by the whole nation of the Jews, was divinely instructed by a holy angel to send for you to come to his house and hear words from you.
Acts 10:22

Chapter Eight: June 2004

The 1st Embedded Trainer Team

The next morning I slept in. I saw Matt at breakfast but I didn't say anything about what had happened. After breakfast I walked across the assembly area in front of the flag poles and entered the headquarters building. I found the Chief of Staff's office and told a clerk that I needed to see him. After about 10 minutes the clerk told me to go in.

I stood in front of COL A and saluted. He returned my salute and told me to sit down. He had some papers in front of him which he glanced at after I sat down; then he smiled at me and told me that I was being transferred to the 1st Embedded Trainer Team at Camp Julian. Then he said, "They have the best food in the Kabul area. Everyone who can, drives there for lunch; you're going to love it there."

He stopped speaking and when I looked at him, he was smiling in a way that let me know he was expecting a big thanks for sending me to a place that had great chow, like I had come to Afghanistan on a gourmet safari or something. "Will I be getting orders transferring me from TAG to the ETT?" I asked him. That seemed to take up the remaining Oklahoma goodwill he possessed and he told he'd see, dismissing me with a salute. "Check with operations for a link-up with your transportation to Camp Julian. You'll be leaving tonight after the BUB." This was my introduction to my next not so excellent adventure in Afghanistan.

A Season of Discontent

I went back to my tent. The guys were just finishing Dr. Phil, my least favorite morning activity. I wondered if they had American television at Camp Julian, a Canadian camp. The other guys saw me come in and asked what I was going to do at Camp Phoenix since they knew I wouldn't be returning to KMTC. I told them that I was transferring to Camp Julian and they began to congratulate me; I guessed they heard about the food there but I quickly realized that they hated it here and would rather be anywhere else.

I hadn't paid much attention to the signs of NCO discontent because I had been so busy with my duties. This would boil over a few weeks after I left. We were at the six month mark in our deployment and the thrill was gone for most of these guys. The CO and the CSM hadn't done much to build or maintain morale, partly because of the demands on their own time and partly because they were of similar personalities; neither one was very outgoing, and the CO was sometimes downright depressing.

Colonel Knight could have been a good counterweight but he had lost perspective as an officer and was more concerned with his own agenda than with the welfare of the troops. After he was fired, LTC Kevin from RI was moved into the deputy position but the battle lines had been drawn and there wasn't much he could do other than maintain what order and discipline were left so that the unit could complete its mission and return home safely. I said goodbye to my tent mates as they left for the morning drive to KMTC and started to pack.

Moving On

After two hours I was ready to go. I had four duffel bags, a wooden box for my printer, my guitar and my table. I also had a cardboard box full of various components of MREs that I had gotten from the DFAC and stored for contingency purposes along with a large number of Gatorade bottles. These were part of my preparation for the siege that Colonel Knight's fantasy had led me to believe would happen. Knowing the history of the British army's retreat from Kabul in 1842 and worried that Colonel Knight's BS might be true had caused me to pack-rat supplies just in case.

The operations people had given me the name of a major who would be coming to Camp Phoenix that afternoon. At the appointed time I went up to the HQ and found the major. I introduced myself and he drove his WFR down to my tent and helped me load my things in the back. We were joined by a captain from the medical service corps and after the BUB we drove to Camp Julian at the other end of Kabul.

This was the first time I had driven through downtown Kabul. It was a madhouse. Other than the general practice of driving on the right side of the road and the practical rule of the largest vehicle having the right of right of way, there were no rules. The major wore aviator wings; if he flew like he drove, he must have been a gunship pilot. He laid his M9 pistol on his lap and pushed aggressively through the traffic. Several times he went up on the median to get around some vehicle.

When we got to the center of Kabul we came to a choke point where several large avenues came together. The traffic came to a stop and there was nothing anyone could do except wait and watch the surroundings. Street vendors moved through the lines of vehicles selling things and the beggars made their way presenting their retched condition with their outstretched hands and sorrowful faces. I took pictures of whatever caught my eye while watching our surroundings for threats.

When we got through the choke point we took off again careening through traffic. We crossed the Kabul River and headed south toward Dar-ul-aman, the location of Camp Julian. The buildings started to look like old west ruins as we continued south. This was the part of Kabul that was almost completely destroyed by Hekmatyar's rockets[102] when he was fighting against the Rabbani government before the Taliban ran them all off. In the distance I could see the Tejbeg Palace on a hilltop. Camp Julian was just below. After we passed the shot up and burned out 'King's Palace' we drove through the security checkpoint manned by Canadian Army soldiers and continued slowly through the camp. The American section was at the far end, close to the large open area below the Tejbeg or Queen's Palace. From the street in front of our area I had a good view of the site of the opening shots of the Afghan-Russian War.

The Bloody Palace

In December 1979 the Tejbeg was the home of President Hafizullah Amin. He had done away with the first communist ruler Taraki and was now anticipating some help from the Soviet Union because the Afghan socialist revolution was going sour. The Russians on the other hand had decided to intervene in Afghanistan but would have to do away with Amin. The Soviet Union's 40th Army began entering Afghanistan by air and land on December 25th, 1979. Soviet special operations forces were tasked with a number of missions in Kabul, the main one being the assassination of Amin. Like many special operations conducted by green troops, this one would not be a shining moment in Spetsnaz history.

Thanks to compartmentalization, on the evening of December 27th, 1979, when Amin was celebrating the arrival of the Red Army, his Russian cooks poisoned him and his entire family. His Russian doctors, not read in on the plot, did everything they could to save him and succeeded. The Afghan security officers alerted their forces around the palace and reinforced them. A small Russian commando team drove onto a parade ground where a guard battalion was being issued weapons and captured the Afghan officers. Shooting broke out and the fight was on.

Russian paratroopers began to hose down the palace with fire from two ZSU 23-4 anti-aircraft guns. Now the Russian paratroops and Spetsnaz commandos moved toward the palace. They came under heavy fire from the Afghan security forces and suffered heavy casualties before finally gaining entry to the palace. Moving from room to room they killed anything that moved. They found Amin upstairs and killed him and his two sons, five and eight years old. The Russians suffered heavy casualties including the KGB Spetsnaz commander who walked outside after it was over and was shot by one of his men.[103]

Solitary Hunt For Bin Laden

The 1st ETT had B-Huts for its admin offices and the unit was in tents. There was a row of B-Huts being built for barracks but they weren't ready. The tents were not the Force Provider Tents like we had at Camp Phoenix, with AC and heating. These were GP medium tents like the army has been using since WWII. I got out of the truck and went to the orderly room. There was a sergeant monitoring the radio and after introducing myself as a new guy he told me to move my things to a tent beside the orderly room. I waited for some further instructions, like who to report to, but he turned back to his solitaire game on the computer.

Solitaire was the time waster of choice in the 1st ETT and some of the HQ troops were experts. It was as if the hunt for bin Laden and solitaire had somehow been associated. The troops that were not directly involved on the ground were lending their will to the effort by playing solitaire on any available computer with an energy I had not seen since the two year marathon solitaire game conducted by a Marine Major who was the liaison from JTF-6 to the Texas CD Task Force.

Oklahoma Hospitality

I went outside, grabbed a couple of duffel bags and moved them to the tent. The place was crowded with bunks. There was a captain stretched out on a bunk nearby so I asked him if there was a bunk I could use. He got up and directed me to lift one end of the double decker bunk in the middle of the floor and move it to one side. In the process of doing this he managed to get his fingers pinched and let me know that it was my fault thereby giving him a reason to dislike me, a good beginning with my new Oklahoma comrades.

I moved the rest of my gear into the tent and stepped outside for a look around. There was a man wearing DCU pants and a t-shirt throwing a rope over a set of horns, practicing steer roping, a rodeo event popular in Texas and apparently in Oklahoma too. I was to find out that the cowboy culture was possibly even stronger in Oklahoma than it was in Texas. The guy throwing the rope looked at me like he suspected me of being the phantom burglar and after an effort on my part to engage him in conversation I gave up and walked up the street to the front door of the orderly room.

Canadian Rules of Engagement

I stuck my head inside and asked the sergeant where the PX and mess hall were and he pointed to my left. I looked that way and saw a row of buildings across the parking lot and a small road so I thanked him and turned to walk that way. "Wait a minute," he shouted, "you can't wear your pistol inside the camp." Huh. "Go turn it in with Major ___," the man who had driven me here. He pointed to a tent nearby and went back to his solitaire game. I found the Major and he told me that we had to hide our weapons, preferably in a locked box near our bunk when we were in camp; Canadian rules. He took my pistol and told me he would issue me some different weapons the next day so I could turn my pistol in back at Camp Phoenix.

Camp Julian

I went back to my camp exploring and walked in the direction of the mess halls. They were fairly easy to identify so I continued past them in search of the PX. In front of me was a large open field that was probably used for a parade ground and people were walking across it toward a large building on the other side. As they walked they were saluting toward a flag pole and what looked like a monument. The Canadian flag was flying on the pole so I wasn't going to salute it but I did want to see what the monument was for so I walked over to it.

There were four photos engraved in stone representing four members of the Princess Patricia's Canadian Light Infantry killed by a U.S. Air Force F-16 pilot who mistook their target practice on a range for Taliban anti-aircraft guns and dropped a smart bomb on them, in this case the bomb was smarter than the pilot.[104] Three others who had been killed in combat actions were also memorialized and the purpose of the monument was engraved on plaques attached to concrete pedestals in both English and French, reminding us of the dual language situation in Canada. I didn't want to anger the Canadians by not saluting as I crossed in front of their flag so I decided to try and memorize the names of these four soldiers and salute their memory every time I crossed. I saluted the memorial, the least I could do to show remorse for the actions of my countrymen.

I continued on and found the PX. After checking out their offerings I continued my scouting expedition and found the Internet Café, a place where you could wait in line for a free computer and check your email. After having my own line on a high speed internet connection at KMTC this was going to take some getting used to. I checked my email and wrote a note to my wife telling her what happened and where I was now. I told her I would call her on the sat phone during our commo window.

Kabul Street Scene
(author's photo)

The WFR (White Ford Ranger)
(author's photo)

Camp Julian
(author's photo)

Camp Julian from Queen's Palace
(Photo by Major Cowan 1st ETT)

Canadian Cuisine

By the time I finished with my email it was time to eat. Since the 45th Chief of Staff had made such a big deal about the chow at Camp Julian, I decided to debunk that myth and prepared to eat some typical army chow. The Canadians were on a tighter budget than we were so I knew their food wouldn't be any better. Was I wrong! The Canadians were living like kings. It was steak and lobster night so you had to walk outside and pick up your steak from the bar-b-que and bring it inside where you got your lobster tail and whatever else suited you from their magnificent buffet. Instead of Gatorade they had large cartons of fruit juice. There was fresh ice cream and lots of other deserts. This place could get you fat in a short amount of time.

I took my food and found an empty table. All around me were Canadian soldiers. I hoped to meet some of them but now I just wanted to relax and have a quiet meal; but that was not to be. A solder wearing DCUs walked up to the table and asked if he could join me. He was a younger man, a Specialist, and though I had hoped for a quiet meal I thought that this guy could fill me in on the 1st ETT so I invited him to sit down. He seemed like he wanted to talk and I was happy to let him while I chewed on my Canadian steak and lobster.

The Unhappy Soldier

The Specialist was a clerk in the headquarters and the stress of the deployment had worn down whatever goodwill he had ever had toward the Afghans. Since they were now my favorite overseas people, the Afghans had replaced the Colombians, who were the most interesting people I had ever lived with up to this point; this excluded, of course, the Vietnamese, who while interesting, had been trying to kill me and who I never really lived with. I listened with slowly growing anger to his litany of complaints. When I asked, he told me he was a clerk in the headquarters. His hatred for the Afghans was based upon stories he had heard from members of his unit who had complained about the inefficiency and corruption they had encountered while working with the ANA in the field.

I was busy eating so the Specialist assumed that he had a sympathetic ear and continued to rattle on about the treacherous Afghans. I kept my eye on him while he talked and began to suspect that he was the product of some Oklahoma inbreeding that the guys from Texas would mention occasionally after some dealing with the 45th Brigade staff. I was more interested in my dessert than I was in what this idiot was saying but I still hoped to elicit some basic information about the unit so I continued to listen to his attack on the Afghans which had turned into a rant. I looked around and saw that several of the Canadian soldiers were listening to this American ambassador of good will so I held up my hand and asked him to stop talking for a minute. "Everybody is listening, can you hold your voice down a little," I asked. He looked around, his face red and then, turning back to me he let me know loudly how he felt about our allies.

I held up my hand again and began to reason with him. "We are advisors to the ANA. This is like an SF mission. We have to empathize with these people, not condemn them because they have a different culture," sure that my voice of reason would calm his troubled soul. Actually, my words had the opposite effect and he went off about the blankety blank blank Afghans again. With this I got mad and told him, "At Ease!" He tried to say something else so I explained the meaning of 'at ease' by telling him to STFU which he understood.

"I don't know what your problem is, Specialist, but I am ordering you to keep your negative opinions about our mission to yourself when you leave our area and come to a place like this, got it?" He gave me a look of unmitigated hate so I stared into his eyes and asked again, "got it?" He nodded, so then I said, "Let's go." We put away our trays and left the DFAC. As we walked across the parking lot I tried to soften the blow of my earlier words but it was too late; I had made another enemy at the unit and it was the first day. I went back to the tent and lay down. It was a lot warmer in these tents with no AC so I hung my mosquito net and prepared for a night's sleep.

Loading For Bear

The next day I went to see the Major who gave me back my pistol and issued me another M9, an M4 carbine with a number of add-ons, ammunition, magazines, and a PVS-7 night vision goggle. I took the weapons and NODs to my tent and configured them. The M4 was tricked out with an ACOG sight, Picatinny Rail with a 'gangster grip' handle and an infrared aiming device. It had a sling that took me a while to figure out but when I got the weapon configured it looked cool. The ETT was short of pistol ammo also but I managed to fill three magazines. Unfortunately I had lost my interest in weapons so all the neat stuff received little attention from me other than the occasional cleaning.

I got a ride with one of the unit members going to Camp Phoenix. We had to travel with at least two vehicles, four people, and two long guns so they were happy that I was available. When we got to Phoenix I was told that we would leave in two hours which gave me enough time to take care of my business. I went to supply and turned in my pistol and bullets. I said goodbye to the supply people and walked over to my old tent.

There had already been two departures from this tent, men who were injured in accidents and sent back to the states. Now with me gone the remaining men had even more room and now they had moved the bunks around so that they could have a TV room, no doubt so that they could focus even more on Dr. Phil's words of wisdom each morning before they went to work. I was glad that we didn't have TV at Camp Julian. My sign '11 A-Gateway to the USA' was still on the door. I gave the tent one last glance and returned to where the vehicle was parked and waited to return to Camp Julian.

The 1st ETT

After a typical 'Mr. Toad's Wild Ride' through the streets of Kabul, we pulled into Camp Julian in time for lunch. I followed the others to the DFAC and had some more delicious Canadian chow. I talked with one of the guys I went to Phoenix with, a sergeant who worked in operations. He was somewhat friendlier and told me a little about the unit. The CO a LTC HS, was the advisor to an ANA Brigade Commander who was stationed at the Palace, the home of President Karzai, in downtown Kabul. The Palace was also called the Arg, traditional home of the Afghan kings.[105]

Every day the CO and members of his staff would go to the palace and meet with their ANA counterparts, just like we did at KMTC. There were five teams of advisors, including a Canadian team, assigned to the five Kandaks of this brigade. The Canadian team was part of ISAF and as such, was not allowed to leave Kabul. This aggravated them because they were good soldiers and wanted to fight as much as anyone but their hands were tied by politics. Most of the advisory teams were out of town with their Kandaks and the people at Camp Julian were either members of the HQ or team members who were for one reason or another not with their teams.

So far they hadn't had any casualties except for a sergeant major with one of the teams whose ears were ringing pretty loud after either an IED or a mine blew off a fender of his HUMVEE. The sergeant was alluding to the combat that his unit had been engaged in for the past eight months but the casualty figures told me a different story. This is not to denigrate the enormous difficulties some of these men faced in the performance of their duties. We got combat pay during the early years of our effort to rebuild the ANA. It was part of our effort to get Afghanistan back on its feet after so many years of war, but for most of us, it wasn't combat but that was changing.

The teams had gone to every part of Afghanistan: Herat, near the border with Iran, Maimanah, to the northwest in the middle of nowhere, Kunduz, up north, Mazar-i-Sharif, up north, and Gardez, down near the border with Pakistan. It had not been an easy tour. The transition between the forces of the regional warlords and the ANA was a very delicate exercise in carrot and stick diplomacy. The ANA officers would negotiate passage through territory claimed by some warlord and the ETT guys would provide the muscle with the flyby of a B-1 bomber 100 feet above the road block.

Funding operations was a big part of what they did and keeping track of expenses was almost as important as watching for the enemy. Artful justifications of necessary items that cost more than they were allowed to spend were just one of the many new skills a combat advisor had to learn. For example, a much needed generator could be purchased, notionally, as three components, each one under the spending limit and reassembled notionally, and put to work. Most of these skills were invented by the smart guys and learned by the others in the classic old school method of OJT or On the Job Training.

The stress on these guys was high and accounted for a few of the men who were in the camp with no apparent function. No one was talking to them and they were minding their own business. Since that was the kind of treatment I was getting I figured that I was considered another guy who had screwed up somewhere else and was now being sent here. Well that was true in my case, but I wasn't going to lie in my bunk for the next three months and wait to go home.

Construction on the 'B Huts' was completed and the unit began to move out of the tents and into buildings with lights and AC. I helped the operations people move an antenna and the radio equipment and that was my entrance into operations. The Operations Sergeant was a MSG and one the few AGR guys who came on the deployment. We didn't hit it off but we were old so we were polite. I didn't like AGR people and he didn't like Texans.

1st ETT Operations Section—Commo Problems

I offered to fill in when needed in the TOC schedule and found out that there was no schedule. The MSG, the Chief of Operations (CHOPS) hung around during most of the day and the specialist I had charmed in the DFAC worked the night shift. Another sergeant, also AGR, was the assistant operations sergeant and worked days. He was totally involved in mail order devices that guaranteed to increase the length and girth of what God had issued him and whenever he was around this was the subject of conversation.

There was no shortage of communications equipment. The TOC had a telephone connected to the Camp Julian system. There was a SATPHONE that was used by various people to call home each night and an FM radio system that was used for communications with Camp Phoenix and our vehicles. A Harris AM radio with automatic link establishment (ALE) was there for talking to the teams in the field and there were several desk top computers connected to the internet that were mainly being used for 'the hunt for bin Laden.'

I spent the first few days learning how to operate the Harris radio. Since there were no instructions, I wrote a basic SOP to help me remember the procedures in case we really needed to use it. I was amazed with the ALE capability, having struggled with HF commo in the LRSD ten years earlier. We made commo checks with Camp Phoenix several times a day but there was no scheduled contact with the units in the field. I thought that that was strange but when I asked, the Ops sergeant told me not to worry about it.

Whenever a convoy left Camp Julian they were supposed to check out with our TOC and call us when they got to their destination. This SOP was loosely followed. I saw that if someone ran into problems in Kabul there was a good chance that we wouldn't know they were gone or where to look for them if they were to contact the TOC with a cry for help.

I began to investigate what was going on here and it took about a week to figure out that there were some serious problems in the 1st ETT. The ETT HQ didn't command and control its units in the field. These units were now OPCON to other units like the SF. Logistics were handled by the 45th Brigade. The only thing the ETT CO could do to assist his teams was to intercede for them with the 45th Brigade Commanding General when problems too big for them to solve came up. Unfortunately the 45th Brigade staff had poisoned that connection with typical staff activity known as backstabbing in Texas.

The 45th Brigade had pretty much gutted the relationship between the ETT HQ and the deployed teams. Eventually the relationship between the 45th Brigade HQ and the deployed teams could be described by the words contempt and hate. The guys in the field had contempt for the HQ guys as in all wars. The guys in the HQ hated the ones in the field because of their contempt and the problems that they caused for the ones in the rear with the gear. What made it even more interesting was the fact that they were all from the same unit back home in Oklahoma and when they went home they were going to have to work with each other again.

The 45th Brigade's CG had recently visited a team that was actually getting incoming rockets at their FOB and they had had some contacts with the Taliban. When the General popped in he found them all wearing beards like the SF guys they were working with and ordered them to shave. This made the field guys lose face with their ANA counterparts which contributed to the hate factor which was now infecting the ETT.

The SATPHONE was interesting because it was the one communications device that would allow the ETT HQ to talk to the teams in the field without 45th Brigade monitoring their conversations. The problem was that the convoluted SATPHONE contracts allowed some phones to call others but wouldn't accept calls from phones using a separate contract. The phone in our TOC could call the teams in the field but the teams in the field couldn't call our TOC.

The commo officer for the ETT was the one whose fingers I pinched when he helped me move the bunks my first day with the ETT. When I approached him about the problem with the SATPHONE he started out defensive and quickly went into the attack mode based on his rank. I tried to assure him that I was trying to help, not criticize him or the commo section but he stormed away in anger.

A Kindred Soul

That afternoon a group of men returned from Maimanah where they had been supporting a Kandak based at the airfield. One of them was a very big black First Sergeant whose bunk I had moved when I moved into the tent. His name was O and he was a little pissed about the new arrangement. I introduced myself after he came back from chow and he said, "I heard about you." A great introduction I thought and asked him what he had heard. "That you screwed up where you were, he said, "and they sent you here; the same thing that happened to me."

I offered him my hand and we shook. His grip was strong and I could tell that he had been spending his spare time in the gym. He smiled as he shook my hand so I figured we could be friends and soon we were. He had been the First Sergeant of one of the companies that pulled guard duty at Camp Phoenix and had run a-foul of someone at Brigade and was sent here for his sins. He had just returned from the field where he had been in charge of a logistics operation at the Maimanah airfield.

The first thing I asked him was how he got an assignment in the field. He just smiled and told me to be patient. "I'm an old man and I'd like to get out one more time before they make me turn my stuff in, if you hear of something please let me know and if anyone asks you, I want to go," I said. Then I declared, "this place is strange, if you know what I mean." His smile told me that he knew exactly what I meant and from then on we worked together to fix some of the things that were broken.[106]

I decided to work on the SATPHONE. If the one in our TOC could call the teams in the field but they couldn't call us I reasoned that they wouldn't be happy having to monitor it 24/7 waiting for a call from the HQ and not being able to call if they needed something or just wanted to talk to the CO. I mentioned this problem to O and he said that was just the tip of the iceberg. There was, he said, a commo warrant officer at Camp Phoenix who might be able to help so when the guys went to the BUB that night I rode with them to see if I could find the commo chief.

The Battle Update Briefing—The BUB

The BUB or Battle Update Briefing was a new name for me. When we first arrived I thought it was an Oklahoma thing, a gathering of the 'bubbas' for an end of the day palaver. The BUB was what the army called the nightly commanders briefing about the activities in his unit during the past 24 hours. This ritual consumed several hours of work by the operations officer of each subordinate unit of the 45th Brigade and our coalition partners. Each unit had to have their power point slides in the computer in the Brigade briefing room at least fifteen minutes before the BUB started.

All the unit commanders would be seated at the U shaped table, talking and joking quietly until the CSM called the room to attention for the entrance of the General. Then the briefing began with each unit commander going through his slides and emphasizing his successes while minimizing his failures. The General was on top of things and could ask some hard questions when he wanted, but if everything was going smoothly and there had been no incidents he managed to finish the meeting in an hour so that everyone could get over to the DFAC before it closed.

I had already been to a few BUBs and since I wanted to stay out of the TAG CO's presence, I scouted around the backside of the HQ until I found the commo shop. The warrant officer was there so I introduced myself and explained the problem we were having with our SATPHONE. He knew exactly what I was talking about and explained the whole multiple contracts issue that had caused the problem. He was a little disturbed that we hadn't come to him before this but I told him that the unit had been busy in the field. He told me to bring the phone in the next time I came to Camp Phoenix and he would switch it out with one that worked both ways. Savoring my success I walked over to the PX parking area where I met the guys after the BUB and we headed back to Camp Julian and some Canadian chow.

My New CO

The next day the 1st ETT commander showed up. He and some of his staff had just returned from a trip to the field where they had set up security for a visit from President Karzai to a provincial capital near Kabul. The trip had been a rough one, using old five ton trucks donated by the U.S. from our Vietnam War stocks. They had a lot of breakdowns and everyone was smoked. The CO was a LTC from the Oklahoma National Guard named HS. He was blond haired and balding with a ready smile and a twinkle in his eye. He was an infantry officer with airborne and air assault wings on his DCUs and he looked fit. Someone introduced me to him and he, unlike some of his subordinates, welcomed me to his unit.

I liked him from the start. He had a way about him that was magnetic and when I found out he was an Oklahoma congressman in his full time job I shouldn't have been surprised. I had served with a Texas congressman who was my CO in the LRSD who had a similar personality and now I was serving with another one. Usually politicians who serve in the National Guard and Army Reserve are a pain in the rear for those above and below them so I decided to keep my mouth shut with HS until I got to know him.

New Quarters

The next morning the CSM told me and First Sergeant O to move our things to the 'B Hut' behind the HQ's building. We helped each other move our bunks to the new building and then we moved our clothing and equipment. I took a space next to the door and O set up across the hall. The CSM moved into the other end of the building along with CHOPS. The other ops sergeant and my favorite specialist moved in along with a mysterious major and a couple of others who worked in the HQ. Someone turned on the mini-split AC and we started living large.

Coalition Warfare

Everyone was happy with our move to the new buildings except all of our coalition partners living at Camp Julian generally, and specifically the Canadian Regimental Sergeant Major. He was checking out the tents that we were moving out of and found a couple of Russian machine guns that one of the teams had brought back from some weapons cache that they found earlier in the deployment. The Canadians were very specific about weapons security and when he found the machine guns lying around he went off on our CSM. The CO got a blast from the Canadian camp commander and we moved to the top of the Canadian shitlist, replacing the Belgians who had done some bad thing a while back. The fact that we were the only ones living in wooden buildings with AC didn't help since the rest of the camp was living in tents and summer was here.

There was also the problem with the internet. When the 1st ETT had come to Camp Julian someone had promised the Canadian commander that the Americans would provide a high speed satellite internet connection like the one we had at KMTC. Never make any promises you can't keep was one of MSG Bil's rules of warfare and it would have been good if the one making these promises had been schooled in the gentle art of special operations like I had.

Soon after the unsecured machine gun incident, I went to the Internet Cafe to check my email and found myself staring at a sign on the door, 'No Americans!- By Order of the Camp Commander.' That was embarrassing. When I turned to walk away I caught several German soldiers grinning at me, revenge at last for the two insults to their warrior nation administered by my forefathers.

Bad Things at the Chopper Pad

The next event to increase the animosity between us and our Canadian brothers occurred when we decided to conduct reinforcement and resupply operation on the large helipad in front of the Queen's Palace. We were sending about 100 ANA soldiers and some supplies on two CH-47 Chinook helicopters to an FOB on the border. We made the arrangements with the aviation unit that was supporting us and at the appointed place and time they came in. I was on the spot with my video camera and I noticed a strange shadow come onto the screen behind one of the helicopters as it landed.

It was an ERO so the Afghan troops were hustled aboard along with a major who had been in Afghanistan for two years now and had extended again. He was a field soldier with the usual disdain for the HQ pogues and we had crossed swords over some administrative issue and had come to an agreement to stay out of each other's way. He was keeping his part of the deal by returning to the team in Gardez and I wished I was going with him so I could show him that I was a field soldier too.

When the helicopters were loaded they took off and headed south, escorted by a pair of Apache attack helicopters. When the dust storm dissipated I saw what had caused the shadow on my screen. A large amount of PSP or perforated steel planking had been ripped up by the rotor wash of the huge helicopters and was now folded in awkward sections on the pad. The Canadians were furious.

Apparently the possibility of this occurring had been raised when we requested permission to use the area rather than do it at the ANA camp next to Camp Julian. One of our officer's poo poo'd their worries but now we had destroyed a good part of their nice chopper pad that was never used again by us while I was there. We said we were sorry and promised not to do it again and the Canadians accepted our apologies after we fixed most of the damage.

The 'French' Canadians

I was working in operations one morning when one of the Canadian captains came into the working area in the back of the TOC. "Bon jour, mon Capitan," I said, the extent of my knowledge of French now fully utilized, but it was enough. A big smile broke out on his face and he said, "I wish those other Canadians would talk to me like that!" We were friends from then on and I gathered from his remark that the Western and Eastern Canadians were not that friendly.

A Western Canadian unit was conducting a RIP with the French Canadian unit that had become somewhat hostile towards us after the machine guns and the internet problems. Everyone hoped we could get a fresh start with the new unit and we did. LTC HS went over and visited with the new Camp Commander and gave him some of that Oklahoma political charm. The next day the sign was gone from the Internet Cafe.

Now that the command group was back the routine changed slightly but most of the people that didn't have a specific mission made themselves scarce by going to the gym. I was getting more involved with operations in the TOC. We had had two incidents where we lost power when our generator ran out of fuel. There was information available that told us how long a tank of fuel lasted but no one was paying any attention. The first sergeant and I made some spaces on one of the dry erase boards, kept track both of when we filled it and when it needed to be filled again. This cut down on power outages but we still had two more when the night shift ignored the obvious.

Trying To Get Into the War

The team down near Gardez was getting rocketed occasionally and I heard that they were going to send some more people down there. I had started riding shotgun with LTC HS when he went to the BUB. He was a friendly guy and we talked quite often. He told me that he knew why I had been transferred. He was a Christian and he had no problem with what I had tried to do. I enjoyed riding with him because he was the best driver I had ridden with since I got here. Having a LTC as the driver was different and I could tell that he was a real soldier and an operator.

I floated the idea of going down to Gardez to help in some way but he didn't jump on that idea so I backed off and hoped there would be another opportunity. After a few trips he told me to drive back from the BUB one evening and he schooled me on Kabul driving. Any hesitation on my part in tight situations brought a rebuke from him, "Go, go, go!" he would shout. "Don't stop, keep moving," he continued to shout until I began driving like he did. Thanks to him I was 'Kabul Qualified' and I consider that to be one of my top achievements in Afghanistan. LTC HS had high standards and when I managed to meet them I felt better about myself in my new unit.

Unfortunately, not everyone in his command shared his standards. When he relieved a non-performer earlier in the deployment and sent him back to the 45th, he wasn't given a replacement. He must have done the math and realized he was going to have to get what he could out of his subordinates. He had a headquarters staff with two majors; one was his deputy and the other was his Ops officer, along with several captains and a number of senior NCOs. There were five embedded trainer teams, each one commanded by a major with captains and senior NCO's.

Some of the men had gone ROAD, retired on active duty, and rather than have them shot like some of them deserved he did the National Guard shuffle that I had seen so many times before and moved the non-performers back to the HQ and replaced them with whoever was still in the game. We were coming to the end of the deployment so if someone tanked now they weren't replaced and the team struggled on shorthanded. I enjoyed riding and talking with him so if he didn't want to send me to Gardez that was his business, but I did let him know periodically that I wanted to do more for the war effort than eat great food at Camp Julian.

Meanwhile—Back At the TAG...

I ran into a few of the TAG troops at the PX when I went over to Camp Phoenix on the BUB run. Major P, a good man, came up to me and congratulated me for escaping the TAG and getting assigned to an ETT. He had had a run-in with the CO over a parrot that he had inherited from the Vermont troops and had become quite fond of. The word came down that all pets had to go away so he initiated the complicated process of getting the parrot sent back to the states. When the CO found out he ordered the Major to cease all such efforts even though the Major's plan was feasible and wouldn't cost the army any money or any of the Major's on duty time. That was a crushing blow to the Major and served to enrage the NCOs who had their own grievances with the CO.

The CSM must have informed the CO that trouble was brewing in the NCO ranks because the CO called a meeting and asked for their frank opinions on what the problems were and what, if any, solutions they could offer. Carl, the sergeant that used to ride with me to work each morning, our IT chief, fell for this and complained that he had seen the CO and the CSM driving around in a single vehicle movement without a long gun and everyone knew that was against the rules.

The CO had a quick solution for that problem and made the good sergeant his security for the rest of the deployment. Later when I saw him in the PX, I asked him how he was enjoying riding with the CO every day. He just glared at me and didn't say a thing. This behavior I would expect if I had had something to do with his difficulties but his bitterness about his plight knew no bounds. The rest of the NCOs bit the bullet and assured the CO that it was all good. Their lack of moral courage made the CO's new security even more furious because no one was backing him up and now he was the only one on the CO's shitlist and the morale in the unit dropped even lower.

A Rumor of War

At the end of June I started to hear talk about Mazar-i-Sharif. It is the largest city in Northern Afghanistan and we had a team there. There were plans being made for the CO to visit and when I heard about it I raised my hand again and asked if I could help. The S2 asked me to gather the map sheets for a road trip between Kabul and Mazar-i-Sharif, a journey of about 280 miles across the Hindu Kush mountain range to a place near Balkh, an ancient city of Central Asia. If I couldn't get down to Gardez, I might be able to have an excellent cross-country journey up north.

I gathered the map sheets; there were quite a few at 1:50,000 scale and put them in order. Then I got a second set just in case. My work was observed and I was told that I might be going along. I still had Falcon View on my laptop so that gave me a backup if the other maps got lost. Our civilian GPSs worked ok in Afghanistan but the mapping software was limited to the large cities and major roads. With them we could get grid coordinates and with the maps I was confident that we could navigate well enough to find Mazar-i-Sharif. We would have our SATPHONE and I had my own as a backup. We couldn't get another HF or a SATCOM radio but the SATPHONE had proven reliable so I wasn't worried about commo.

I had been working on the vehicles and initiated a scrounging operation to acquire some tow straps. There were a large number of air force aluminum pallets laying around near the PX at Camp Phoenix and I had started grabbing a couple of tie down straps from them every time I went to the BUB. I had just about picked up enough for a tow strap in each vehicle. LTC HS saw me filch a couple one night after the BUB and appeared worried. I laughed at his fears that I might get caught but I could tell that he was just trying to avoid another ass chewing from his friends at Brigade.

I still wasn't sure that I was going on this trip, so on the way back from the BUB one evening I tried to sweeten my request to go by offering to make a video of the trip that would be useful in the years to come to tell the 1st ETT story. I don't know if that had any effect on the decision making process but the next morning I was told to pack my bags for Mazar.

I got on-line and found a sale on AVIOs, a proprietary video editing machine I was familiar with, at a company in Austin. I emailed my wife and asked her to buy one and ship it to Camp Phoenix along with a monitor. The AVIO was the size of a large VCR and the monitor was about twelve inches square. She sent these to me by mail from the U.S. for about $30, a modern war perk. They would be here by the time we got back from Mazar so I could make good on my promise.

To Mazar-i-Sharif with the RAF

Trip planning continued and soon we were told that rather than driving we would put our Ford Rangers on a RAF C-130 and fly. That was a disappointment to me because of all the map work and my hopes to see the country but I reasoned that we might have to drive back so we kept the maps. The people going now included the 'CHOPS,' a LTC medical service officer who wore an SF tab, the CO, his favorite interpreter, the commo officer and me. We loaded the vehicles early in the morning and headed for the Kabul airport. As we drove through the gate to the Afghan Air Force side of the airport, there was a Russian MIG 21 on a pedestal near the entrance and as we drove to the RAF ramp we passed a number of MI-8/17 Russian helicopters and some MI-24 Hind gunships. There were a number of wrecked Russian airplanes around the airport which gave testimony to the fighting that had gone on here in the past.

The Brits drove our trucks onto the airplane and tied them down. There was a third truck added with two guys in civilian clothes. One of them looked a lot like LTC HS and as it turned out it was his cousin. The other man was my old acquaintance from G Company days when he was the Regular Army advisor to the LRRP Company. I had seen him at KMTC when we first got there and now we were flying to Mazar together. They were both working for MPRI and were going up to observe the unit our team was advising while they conducted a staff training exercise.

There were a few other passengers: some British soldiers, a guy from the DEA, and some civilians. The Brits were running the Provincial Reconstruction Team in Mazar. Also in Mazar, was a large part of the 1st Battalion, The Green Howards, who had replaced the 2nd Battalion, Royal Gurkha Rifles, part of UK's contribution to ISAF. The RAF C-130 was part of the British logistics effort to keep their people supplied.

After a short safety briefing the crew chief told us to sit down and hang on and we took off for Mazar. I was taking photographs and video as we circled over Kabul, gaining altitude before heading northwest toward the Hindu Kush. I tried to get a glimpse of KMTC but we were too high and I couldn't get oriented before we left the Kabul area and started over the bleak landscape of Afghanistan. I continued to observe the ground as we flew along and saw several places where it looked like there had been some heavy bombing around some outpost in the mountains. I wondered if I was looking at some epic battleground where the Russian soldiers had fought to the death against a mujahideen assault 15 years before or a more recent battle decided by precision air strikes during the U.S. invasion.

We finally crossed the mountains and ahead I could see a small patch of green in the middle of the dull light brown color of the desert below. We began losing altitude so I figured that the green patch was Mazar-i-Sharif. The loadmaster came over and asked me and several others to sit on the back ramp of the plane when we landed. This was a kind of weight redistribution move which the air crew felt might increase our chances of an uneventful landing so I hastened to comply.

Sitting on the back ramp would give me a longer life than the guys up in the front since sometimes the only survivors of a plane crash are the ones sitting in the back. The plane made all the usual noises as it approached the airport. I continued to shoot video out of the porthole closest to the rear ramp until the loadmaster waved his hand at me to sit down about a minute before our wheels touched the ground. The pilot threw the props in reverse and stood on the brakes. We slowed down pretty quickly causing the tail of the plane to rise at the same time, which is why they asked us to sit in the back. The vehicles pulled at their tie down chains but everything held and we were soon taxiing down the runway.

Mazar

Shortly after we touched down I was back on my feet and taking video of the wrecked Russian aircraft and bombed out anti-aircraft positions that lined the runway. The plane turned around and parked on the ramp in front of the control tower. The loadmaster came back and lowered the ramp and we walked off the plane. There were several guys in DCUs and a couple of WFRs standing to one side of the control tower and passenger terminal of the Mazar-i-Sharif International Airport. Along with our plane there was an MI-8 on the ramp and a small twin prop airplane as well. Next to the terminal building was a Russian radio vehicle and large antenna surrounded by a barb-wire fence. The radio truck interested me but by then the Brits had unloaded our vehicles and it was time to move.

The ETT leader, a Major --- saluted LTC HS and welcomed us to Mazar. He told us to follow him to the ANA camp a few miles from there. The LTC was letting me drive now that I was 'Kabul Qualified' and so I kept a decent interval from our host's vehicle and followed him out to the main highway. To the left was Mazar and I was prepared to turn that way when the vehicle in front of us turned right and headed up the highway.

I followed and entered the main road between Mazar-i-Sharif and Kunduz to the East. This whole area had been a battleground during the U.S. invasion and I'm sure it saw some fights during the Jihad. The usual wrecked tanks and other armored vehicles were scattered across the landscape as far as the eye could see. There were camel caravans' way out in the desert and pedestrian traffic close to the road. The traffic was light and everyone seemed to be in a hurry. I kept waiting for us to turn into an army camp but we kept going east farther and farther away from Mazar-i-Sharif.

Mazar-i-Sharif—East Camp

Finally we came to a road intersection. There was a road going north but we continued to the east for another ten miles before turning at last and driving down a dusty dirt road to an ANA camp in the middle of nowhere. We passed a tall berm on our right and when we came to the end there was an entrance. I looked in and saw a large collection of Russian tanks, artillery and other heavy equipment; this was one of the collection points for heavy weapons that were now being turned in by the warlords as the Afghan Transitional Authority began to consolidate its power.

The ANA was guarding this equipment and their camp was just beyond the weapons cantonment area. The camp was made with HESCO barriers just like the ones at KMTC. There was a water tower that could be seen above the walls with an Afghan flag above it waving in the breeze. There were guard posts at each corner of the square shaped fort and an SPG-9 recoilless rifle on the wall facing the highway. As we approached the entrance to the camp an ANA sentry gave a loud shout and the vehicle ahead stopped and assured him that we were ok and he waved us in.

Inside the walls was a crowded area filled with GP medium tents, no luxury living here. There were two mortar pits dug into the ground and well laid out. I was impressed with the neatness and practicality of these positions and with the other defense positions. It looked like the ETT and the ANA had been taking care of business. The troops were beating the heat by lying on their bunks in the shade of their tents with the walls rolled up. The ETT had three tents at one end of the compound. Against the inside of one of the walls was a large shipping container with the words, "Buck and Cutters House of Pain," spray painted on the side. Inside was a bar bell and some free weights.

The first tent was used as a TOC for the ETT. They had a Harris HF radio, a SATCOM radio, an Inmarsat Internet data link and some Motorola handhelds for local traffic along with some vehicle mounted radios using the same frequencies. They kept a map of the area on the wall with a status board for personnel and vehicles, altogether a squared away place. The Major gave us some bunks in the back of that tent and we stored our stuff there before joining the other ETT guys for lunch in the chow tent.

They were eating pretty high on the hog—of course we wouldn't mention that to the Afghans who held differing views on what was OK to eat. There was a U.S. logistics base back at the Mazar-i-Sharif airport that was run by the regular army and the ETT was able to purchase all their food there. They hadn't quite been able to blend in with the ANA to the point of eating the Afghan food and since I was a guest I decided to keep that observation to myself. The ETT guys had the normal level of suspicion about the appearance of their CO and were waiting to see what was going to happen.

The CO was pleased with the appearance of their camp and the ANA's progress so they were polite to me as they tried to figure out if I represented a threat. They all knew that 'CHOPS' was a bundle of BS most of the time and he didn't change gears a bit when he got with the ETT and I figured he would fill them in on my presence.

Loading a RAF C-130
(author's photo)

Flying over Kabul
(author's photo)

Mazar-i-Sharif International Airport
(author's photo)

MIG 21 at Mazar-i-Sharif International Airport
(author photo)

A Walk in the Desert

That afternoon I got bored and after climbing to the top of the water tower for a look-see, I decided to take a walk outside the camp. I asked one of the sergeants about the mine situation and he thought that the area had been cleared so I took a walk. As I moved I looked around first at the ground, before I put my foot on it and then at the area around me. I had my vest with a Camelback, my M4 and pistol; the IBAS I left in the tent.

The land around the camp was basically desert for a long way in every direction except south where the Hindu Kush Mountain Range rose up like a solid wall from the desert floor about twenty miles away. There was a little scrub vegetation here and there in the dust. The wind was blowing hot air across the land and I could see some big dust devils in the distance. The air was so dry that it immediately evaporated any moisture from my skin so it was hard to tell how much water I was losing in the 120 plus degree heat. I kept the Camelback spout close by and sucked a drink every few minutes.

As I got farther from the camp I started to see signs that there had been a military camp here before. There were a few remains of some foundations and a couple of places that looked like they had been basements that were filled in after the buildings had been knocked down and cleared away. I came upon the small turret of a BRDM lying on the ground with an empty machine gun mount. It looked like it had been there for a while so I imagined I was looking at a Russian Army camp from the days of the Jihad.

Then I saw something shining on the ground near the turret and after making sure that it wasn't attached to something deadly, reached down and picked up a silver button with a hammer and sickle on it. There was a piece of cloth attached like it had been torn off a uniform so I wondered if it was there because some Russian soldier, deciding that he had had enough of the BS that usually accompanies every army, ripped it off or whether this was all that was left after some high explosive event. I looked around for more artifacts but that was all I could find. The light was changing, it was getting late, so I walked back in the footprints I had made and entered the camp.

I joined the CO and the ETT in the kitchen tent and listened to the small talk. The plan for the next day was to go the airport in the afternoon and do some business at the ALOC, visit the Brits, and go into

Mazar to the PRT. I walked around the camp one time and turned in. The tent we were sleeping in had a mini-split AC attached and it was straining noisily against the 120 degree heat outside when I finally fell asleep.

I woke up before dawn, about 0430, dressed quietly and went outside for a look around. The camp was quiet. I could see the sentries at their posts on the wall and I waved to the one at the corner near our tent. He waved back so I climbed the ladder and sat next to him. He couldn't speak any English and all I could say was, "salaam alaikum," to which he gave the standard reply so we sat there together in silence looking at the stars and appreciating the special qualities of the predawn darkness all around us.

After about ten minutes I patted him on the back, gave him a breakfast bar from one of the pouches on my vest and climbed down the ladder to the ground. I continued on past the camp entrance and climbed the water tower to look around. Other than the occasional vehicle on the highway all was quiet; so I just sat there and stared out toward the Hindu Kush while calling on the Lord quietly until I saw the beginning of morning nautical twilight (BMNT) the official term for the dawn.

The Old Testament

The ANA was making a morning presence patrol to a village in the foothills of the Hindu Kush south of our fort. I joined the CO and several members of the ETT and we tagged along. After a long ride across the flat dusty plain we began going up into the hills. The village came into sight and the first thing I saw was a woman in a blue Chadori riding on a small donkey sidesaddle, without the saddle. There was no electricity or stores and no cars on the dusty path and I felt myself being transported back to the Old Testament times.

After the children of Israel were carried away: first to Babylon and then to Persia, they were released by Cyrus to return to Jerusalem. Only a few of the Jewish captives returned to Israel and the rest disappeared into the world. There were a number of theories about what happened to them and the one that I subscribe to is that the Pashtun people are some of the wandering Jews from that time. Seeing the woman on the donkey made me think that I could have been looking at Mary, the mother of Jesus, on her way to see Elizabeth.

A Trip to Town

We returned to the camp and I joined the others in the cook shack. The CO told me I would be driving his vehicle in a convoy to town. The ETT commander would drive the lead vehicle, the CHOPS and the commo officer would follow and we would take up the rear. After breakfast we loaded up and took off. Leaving the camp, we passed the heavy weapons area and turned onto the highway. There was more traffic now and we passed anything that wasn't going at least 70.

As we got closer to the airport the traffic made it difficult to pass so we started bunching up. The ETT commander passed a van but CHOPS was not able to. He rode right up behind the van, a move that may have meant something in Oklahoma. However, the van driver was either angry at the tailgating or reacting to something we couldn't see. He put on his breaks and CHOPS ran into the back of the van. I had gotten too close to CHOPS and knew I was not going to be able to stop so I swerved to the right and slid to a stop on the dirt shoulder. What I hadn't noticed was the thirty foot drop-off about a foot to the right of my right wheels. LTC HS looked out the window to his right at the drop-off and then coolly turned to me and said with a smile, "Nice move," a compliment indeed from an LTC who could see we were 12 inches away from death or serious bodily injury.

I could see the drop-off now and while I was acting like it was all part of the plan, I told the CO to "Praise the Lord!" and he replied with, "Amen!" The CHOPS backed up and went around the van, breaking his mirror in the process. The CO told me to follow him and I did. The van didn't look badly damaged but we didn't stop to exchange driver's licenses and insurance information with the Afghan driver who was now sure to be susceptible to our enemy's propaganda after his encounter with the Americans.

Kandak HQ—The Airport

We turned off the highway at the airport entrance and drove down the long road. At the actual entrance to the airport we turned left and drove into a small area where the ANA had a Brigade (Forward) headquarters at Mazar-i-Sharif. The CO and the ETT commander had a meeting with the Brigade commander and I walked around outside. There was a group of ANA soldiers standing around so I walked over and stood with them. They were all checking me out as I greeted them.

One of them went inside and came out with one of the interpreters; they wanted to ask me some questions. How old was I? Did I like Afghanistan? Was I married? All questions I had come to expect from the younger soldiers. The next one surprised me when one of the older soldiers asked me if I had a lot of sex with my wife, making some quick gestures that caused the others to break out in giggles. I thought of what reply I could give them and then told them that we shouldn't talk about things like that which surprised them but they all made sounds like they agreed and the conversation went elsewhere. They let me take some pictures and the interpreter took some pictures of us in a group. They wanted me to hold an RPG for a hooah picture so I was doing so just as the CO came out.

The ALOC

He told me to get in and drive; we were going to the ALOC site. On the way he told me that the meeting went well and that we would be going to the Qala-i-Jangi the next day. That was the place where Mike Spann, the CIA officer, had been killed in the prisoner revolt during the opening days of the invasion. Johnny Walker Lindh, the American Taliban, had been captured there after the revolt was terminated with air strikes and heavy fighting.

As we drove to the ALOC we passed the Jordanian Hospital, a humanitarian relief effort of the Jordanian government. They had Jordanian Army soldiers pulling security and a Jordanian flag flew over their compound. The American ALOC site was guarded by MPs from the same unit that guarded the Alamo at KMTC. They waved us in. We parked and went inside to a dining area that served as a lounge for people passing through.

The main attraction at the ALOC was the female captain in charge. She and her unit were with the 25th Infantry Division and had been running this site for long enough for the word to get out all over Northern Afghanistan that there was a beautiful army captain at the Mazar-i-Sharif airport. Her presence accounted for a lot more activity there than there would have been if I was running the place. She came in and welcomed us with an amused smile. She was not unaware of the effect that she had on the guys and she bore it with a friendly professionalism that took the bite out of her rejection of all the subtle flirtations of the ETT guys who had been coming here for their supplies.

The captain's assistants worked with the ETT guys on their food orders while the CO talked with the captain and expressed his appreciation for the way that she and her unit had supported the ETT. He told her that we would be visiting the Qala-i-Jangi fortress the next day and offered to escort her and the members of her unit who wanted to get off the site and see a little of the country. She accepted the invitation and after checking with the NCOIC, told him that seven of them would join us. The CO told her to be ready at 0900 the next morning.

We ate lunch at the ALOC and then moved on to our next task. This was to rent some vehicles to be used for transporting a team of U.S. Army Corps of Engineer civilians out to a location near Mazar-i-Sharif where the HQ of the Northern Corps was to be built. It was part of the next step in building the ANA. The Central Corps had been established in the Kabul area and now there would be five ANA Corps around the country. The Kandak at Mazar would provide security for the construction site and we were tasked with providing transportation and security for the visiting engineers.

Business in Mazar

We drove into Mazar; the CO was letting me drive so I knew he hadn't lost confidence in my driving. We followed the ETT commander to a place on the east side of Mazar where there was a vehicle rental. The operator of the rental company came out and stood with the CO and members of the ETT. This was not a simple task and the CO's interpreter took over the negotiations. He was a doctor who was doing like many of the other Afghan doctors were doing, serving as an interpreter until things stabilized and he was able to earn a living in the medical field.

'Doc' was an entrepreneur and already held the contracts for food service at the two camps supporting the Mazar Kandak. I don't know about conflict of interest rules but I do know that the ETT faced enormous problems in the carrying out of their duties. If someone like 'Doc,' the interpreter, stepped up to the plate and offered a viable solution to problems like feeding the ANA they went with it and were happy if he made some money for his efforts.

I pulled security while the negotiations dragged on but finally they reached an agreement. The ETT major who handled the finances pulled out a wad of Afghanis and counted them out on the hood of the WFR. The 'Doc' took the money and he and the rental operator talked for a moment, the 'Doc's' back was to us so whatever took place as the money passed hands was their business. We now had two SUVs for the engineers when they showed up. The next stop was the PRT.

We drove into Mazar. The Blue Mosque dominates the center of Mazar-i-Sharif. Supposed to be the tomb of the son-in-law of Mohamed, it is an impressive building with many arched doorways, domed roofs, and several minarets. The building has a lot of blue tile giving it its name and there are hundreds of white pigeons flying around it and walking on its grounds. The CO said that we might visit the mosque later when we had the ANA with us but I wasn't that excited about inciting a riot and being torn apart by enraged Muslims protecting their religious places from the infidels.

The PRT

The Provincial Reconstruction Team[107] was in a quiet neighborhood and kept a low profile. The Brits were running it and we were visitors. I don't think it would be easy to compile a complete history of the Afghanistan campaign without a substantial section on the PRTs. The Brits had an effective operation and I could tell that they were expecting the worst from their American cousins as we walked in. I was surprised to see a soldier from the Green Howards that I had met at Camp Phoenix and we spent some time exchanging travel stories while the CO and the ETT commander conducted business in one of the offices.

The PRT was involved in numerous projects in Mazar. One of the most important was keeping the peace between General Mohamed Atta and General Abdul Rashid Dostum, both of whom claimed Mazar as part of their territory. Atta represented the Tajik faction of the Northern Alliance and Dostum was the leader of the Uzbek's Junbish-i-Milli Militia. You could always tell whose party was controlling a section of the city by the large portraits of one of these guys prominently displayed on the streets leading into that section.

Everywhere in the city, groups of armed men riding in the back of pickup trucks sped through the streets with photographs of their leaders displayed in the front window. There were numerous checkpoints on the main roads manned by these groups. They used the tracks from Russian BMPs as speed bumps so that they could tax the Afghans traveling through. They always waved us through but you never knew. The CO complained about these armed groups every time he saw them because the word had gone out from Kabul for all the militias to turn in their weapons and go home. Apparently the word hadn't gotten here yet or the local warlords chose to ignore it. Either way I didn't think it would be prudent to initiate any action that might start another round of fighting with us being the first casualties.

The PRT worried about the same thing and were polite to the suggestions of the Americans but as soon as we left I'm sure they continued with their own program and filed our suggestions in an appropriate place. The CO came out of the meeting after about an hour and we left. The next stop was a carpet store in downtown Mazar where 'Doc' was going to negotiate for a particular carpet that one of the ETT guys had been trying to get at a price he could afford. There was also an electronic store nearby where the ETT commander was trying to get parts for a generator they had bought there a few months before and was now broken.

Making a Joyful Noise

We drove through Mazar and found the two stores just doors apart and the whole group went in to negotiate, leaving me outside by myself to guard the vehicles. I was wearing my IBAS and helmet as per SOP and I was carrying my M4 on a sling around my neck. It was hot, damn hot, hot as the sun, like Robin Williams said in "Good Morning, Vietnam." At least 110 degrees that afternoon, it was like carrying an extra fifty pounds, standing out there watching the vehicles.

Of course while I was watching the vehicles the local Afghans were watching me; everyone who had nothing better to do in that block and across the street was soon standing around me asking for money. I made the international sign for no money but kept a smile on my face and my hands away from my weapons. Everyone seemed friendly so I began to talk to them in English thanking them for inviting us to their lovely country. I told them how happy I was that the

Afghans were such friendly people and my words seemed to please them because they were all smiling at me so I decided to sing to them. I sang a song from Psalm 116 that we sing in the church meetings; "What shall I give unto the Lord for all he has done for me; I'll take the cup of salvation and call upon the name of the Lord." When I finished the first verse I called on the Name of Lord Jesus several times in a normal speaking voice.

The whole group was standing there looking at me and waiting to see what was next. I sang them another, "Just taste and see that the Lord is good," another favorite of mine. Then I sang the first song again and as I was singing I sensed and then saw that the CO had joined me. I finished the song and waited for him to rebuke me but he said, "That was beautiful." I don't think he was directly referring to my voice but he enjoyed the words. He did however think that we had gathered too many people and he asked me to send them away.

Time to Go

Then a young teenage girl with a scarf over her head but not veiled, placed herself in front of me and made signs asking for "dollars." She smiled and began to flirt with me in the international way that females have that causes men to imagine many things. A man wearing a turban stepped toward her and pushed her, telling her something as she moved quickly away. He turned and glared at me and I tried to assure him that I had done nothing to encourage this contact with an Afghan female, but he just glared at me and left.

I noticed that my right hand was resting on the ACOG of my M4 and maybe he noticed the same thing. The rest of the crowd, sensing that the entertainment was over and that there wasn't any money coming from me, began to go about their business. I kept an alert demeanor now thinking that the guy in the turban might come back with some friends who didn't like the Americans. The ETT guys finally finished their business and came out and we headed back to the ANA camp outside of town.

After dinner that night I was ready for bed. There had been a news report on the internet stating that Vice President Fahim Khan had been fired by President Karzai. The CO asked me what I thought about it and I suggested that we set up a guard roster among the Americans. I had wondered why they didn't have one the first night but he decided it wasn't necessary. I mentioned that some Russians had been killed in their bunks by the ANA during the Jihad but he didn't see it as an immediate threat. Standing around in that heat had taken a lot out of me so I went down pretty fast and didn't wake up until 0500 the next morning.

ANA Camp and Mortar Pit
(author's photos)

Russian Tanks
(author's photo)

107 mm Multiple Rocket Launcher and Anti-Aircraft Guns
(authors photo)

Where do wars and fightings among you come from? Are they not from this, from your pleasures that war in your members?
James 4:1

Chapter Nine: July 2004

Fourth of July Fireworks at Camp Phoenix
(An Unfortunate Series of Events)

Meanwhile, back at Camp Phoenix, Toby Keith and Ted Nugent came to Camp Phoenix just before the 4th of July while we were in Mazar. According to Colonel Davis a good time was had by all. The entertainers even invited some of the 45th Brigade guitar slingers to come up on the stage and play their guitars.[108]

After the show, a senior TAG officer suggested a bar-b-que to try and further improve the morale for the 4th of July. The CO approved and the CSM was directed to get some steaks from the DFAC and organize a bar-b-que to be prepared by the officers—mistake number one. Three of the senior sergeants, realizing the inherent weakness of this plan, determined to 'advise' the officers assigned to perform this task. First though they had to make a quick run to KMTC. When they returned they found no officers and no fire in the grill—see mistake number one.

At that time one of the officers showed up and filled the grill with charcoal but there was no lighter fluid. They all went in different directions to find some. Yet another sergeant, watching from afar and wanting to help, went to the Bulgarians and got a bottle of their home-brew, Camp Phoenix being a 'dry' camp and all. He poured it on the charcoal, lit it and left—mistake number two.

The Bulgarian home-brew was a well refined product and burned with no visible flame. One of the senior sergeants and the captain returned, the sergeant with a bottle of mo-gas which he proceeded to squirt on the invisible flames of the Bulgarian home-brew—mistake number three.

The mo-gas immediately flared and the flame following the stream to the bottle caused it to explode, covering the sergeant and the captain in burning mo-gas. They were immediately medevaced to the states. The result of this morale raising exercise was the feeling that the CO blamed the NCOs, who were the target of the exercise, which resulted in even lower morale.[109]

At Fort Sam Houston in San Antonio the two Rhode Islanders were treated to some Texas hospitality. There they linked up with Terry. He broke his leg jumping from a WFR after he finished removing a 'Don't Mess With Texas' bumper sticker from Colonel Knight's truck; something the 45th 'Oklahoma' Brigade had insisted on.[110]

They were also joined by the medical officer who, having taken himself off the list for medals by putting a 9mm hole in one of the clearing barrels, managed to break his ankle playing volleyball and was medevaced. Soon there were five members of TAG at Fort Sam Houston in San Antonio, representing the wounded TAG warriors of the Afghanistan war. The last month in Afghanistan would prove to be the longest for the TAG.

Tourists at War

The next morning we discussed the trip to Qala-i-Jangi which started the ETT guys to speculating about their chances with the beautiful captain at the ALOC site. With this the CO told us to load up and we headed her way. I drove and was feeling comfortable enough on this road that I began to take in the sites nearby as we drove along. There were a lot of graves along the side of the road with poles holding banners with Arabic writing. There had been some fighting in this area during the invasion as evidenced by the various knocked out armored vehicles just off the road. Farther out in the desert I could see several men leading a line of camels on their way somewhere like they had suddenly materialized out of a time warp and were actually on their way to an ancient city like Samarkand on the Silk Road.

This was why I loved Afghanistan. There were so many things that I would never see if the war hadn't taken me out of my routine and sent me half way around the world to a land beyond time. The view to the north was flat desert that went all the way to Russia. As hot as it was I was glad that it wasn't winter because I could imagine that when the northers came it got real cold.

There was a big checkpoint coming into Mazar where the traffic was always backed up. General Atta's face let us know whose territory we were in. They waved us through and I was relieved to see that nothing had changed even though Fahim had been fired the day before. This was a sign that the Afghans had decided to take a break from their continual fighting among themselves. We traveled a few more miles west where we turned into the airport and went to the ALOC site.

The captain had her troops ready but wouldn't be going herself. The disappointment among the ETT men was almost palatable and several tried to persuade her to change her mind, "Hell, Captain, we'll wait for you," one shouted. The CO finally intervened and after some short contingency planning we went to the ANA HQ and picked up two pickup loads of ANA including the Kandak commander and drove out of the airport and on to Mazar.

The House of War

Keeping our convoy together it took a little longer to get us through the city but we managed to get through intact and were now driving through farm country on the southwest side of the city. Ahead I could see a large object that appeared to be rising out of the ground on the left side of the road. As we got closer I could see it was the crenellated wall of the fortress called Qala-i-Jangi, the House of War.

After turning off the main road we followed a road along the base of the wall that led to the entrance of the fort. Two Russian howitzers flanked the entrance and a large road led up to a large gate with a smaller door in the gate for foot traffic. The Kandak commander, the CO, and the ETT commander went up to the gate. Some fellows wearing traditional Afghan clothing and carrying AK47s slung on their shoulders came out and the Kandak commander talked with them for a few minutes before the gate crew opened the gate and we were motioned in.

The Afghans led the group of us down a lane covered by large trees that kept it shaded. We had left our IBAS in the vehicles guarded by ANA soldiers. I wore my vest so I could have my camelback and my snacks that I used to make friends with the ANA. The lane brought us to a path that led to the left toward another tall wall with a gate. There were two long signs, one on each side of the gate. The guides turned left and we walked toward the signs, one written in English and the other one in Arabic script. I assumed that the English writing was the translation of the other sign.

The guides stopped so we could read the history of Qala-i-Jangi. It was originally built in 1889 by 18,000 workers and took twelve years to complete. The sign listed the Afghan government's construction supervisor and some engineers, then fast forwarded a hundred years to 1992 when General Rashid Dostum chose the fort as his HQ after the fall of the Najibullah regime and ordered it to be repaired. Then the Taliban destroyed it; and Dostum had it repaired again after the Taliban were defeated during the invasion. We walked through the gate in the wall and entered the dark and bloody ground of the House of War.

What Happened in November 2001

The following is an account of the fighting there as told to us by the guides and from information available on the internet:[111] In November, 2001 after the fall of Kunduz, a large city to the east of Mazar-i-Sharif, large numbers of fighters surrendered to Dostum's forces. After separating the Afghans from the foreign fighters, mainly Pakistanis, Chechens, Arabs and other Jihadis from Asia and the Middle East, the foreign fighters were loaded into trucks and transported to Dostum's HQ at Qala-i-Jangi. Here they were herded into the section we were now looking at, a large open field surrounded by high walls. There were a few buildings, including one painted pink.

The prisoners were surely anxious about their fate. Dostum's men had not searched them very carefully and one of his senior commanders was killed by a man who blew a grenade while standing next to him. This event resulted in the prisoners all being forced into the small building where they were kept for the night. The next morning they were brought outside to the open field where two CIA officers began to interrogate them, looking for information on the whereabouts of bin Laden. There was a news reporter taking video as they asked a man who looked like he was of European descent to give them information but he wasn't saying anything; this was the American Taliban John Walker Lindh, a young man from Marin County, California.

Shortly after the CIA officers tried to get information from Lindh, a group of prisoners surged on the officers who fired their rifles and pistols into the crowd. One of them was overwhelmed and went down under a wave of men. The other one backed away firing until his weapon was empty. Then he ran to an exit from that section of the fort, forever haunted no doubt by internal and external recriminations. Meanwhile the prisoners had overpowered and killed some of Dostum's men; the rest escaped. The prisoners now had control of half of the fort.

In some of the buildings were large quantities of arms and ammunition which the prisoners then seized. Now they were no longer prisoners and back in the fight. The next few days were filled with shooting and air strikes. The British and American Special Forces that responded to the call for help made by the CIA officer brought hellfire into that place by using precision bombing from aircraft high overhead. At one point a JDAM, a 2000 pound high explosive bomb equipped with a guidance system, landed on the position occupied by the SF men and some of Dostum's troops. A number of Dostum's men were killed and all the SF were injured by the blast. AC-130 Spectre gunships worked the place over during the night and succeeded in destroying the ammunition still stored in the prisoner holding area.

The next morning Dostum's men entered the battle area and began mopping up. The prisoners who survived were now literally holed up in a basement under one of the buildings. It took two more days before the remaining survivors finally surrendered, only because Dostum had the place filled with cold water. More than eighty men came out including Lindh.

After that the war moved on and most of the attention of the world was to the south as the Northern Alliance and their SF/CIA amigos took Kabul and then Kandahar. Dostum had his fort repaired and now it was a military tourist attraction. The guides had been given some money for their work and let us wander around and take pictures.

The Monument to Mike Spann

The damage to the walls of the fort had been repaired and the bomb craters had been filled in. There were a few dead trees; most of their branches were gone; and long brown grass and a few small bushes covered the ground. The pink building was still intact where the prisoners had been housed the first night. Its walls were pock marked with bullet holes. Near the pink building was the entrance to the basement. I looked in but it was dark at the bottom of the stairs and I had no desire to explore. On the other side of the building was a gazebo-like structure with an obelisk monument honoring CIA officer Mike Spann who was killed nearby. We all stood around the monument and listened to our interpreter translate for our guide as he told the story of what happened. When he was done everyone split up to look around.

I wandered around and took some pictures with my digital camera. Every so often I would stop and listen to the hot wind blowing through tree limbs and bushes. I could only imagine what a different place this was on the first day of December, 2001 after the shooting finally stopped. It was estimated that over 500 men were killed in the fighting and some news gatherers said that it was hard to walk without stepping on a body or parts thereof.

I met up with the CO and we had our picture taken together and then he passed the word that it was time to go. We walked back to the main entrance and I walked with the Dostum men who were our guides. One of them bore a resemblance to Dostum and I wondered if he might be a son left here to keep an eye on his father's interests while he was in Kabul. I took their picture and passed my camera around so that they could see themselves, a good practice in that country.

The CO spent some more time with Dostum's commander and the Kandak commander, probably having some Chai and pledging undying friendship until the next outbreak of fighting. I hung out with the ANA men who had guarded our vehicles and showed them the pictures on my camera and gave them some cereal bars from my vest. That made them happy and when I gave them a couple of bottles of water from the back of the truck we were friends forever, also or until…

The Green Howards

The CO came out and we formed up into our convoy order and drove back through town to the airport where we dropped off the ANA commander and the ALOC soldiers. Then we dropped in on the HQ for the Green Howards and the CO had a meeting with their battalion commander. He invited me to join them and while they were talking I noticed that one of the battle streamers attached to the Green Howards' unit flag said "ANZIO." When the conversation flagged, I called the two LTCs' attention to this coincidence of history.

The 1st ETT's call sign was ANZIO from its parent battalion's participation in the operation there that tried to take Rome during the Italian campaign of WWII. Both men were appreciative of their unit's history and the great cooperation between our two countries then. It reminded us of our interest in continuing our cooperation in this war. I took a picture of the two of them and then somehow, lost control of my camera. I didn't notice it before we returned to the ANA camp for the night.

A Busy Schedule

We ate dinner together that night and the CO filled us in on the next few days' events. We had to meet the visiting Corps of Engineers at the airport the next day around 1300. In the morning we were to pick up the SUVs we had rented a few days before and link up with the engineers. Together we would travel with a security element from the Kandak troops stationed at the airport to the site of the Corps HQ that would be constructed not far from the Qala-i-Jangi fortress we had just visited.

The day after that there would be a Command Post Exercise (CPX) for the ANA Brigade and Kandak officers. This would take place at the airport. After that we would travel to the other ANA camp on the west side of Mazar where the Kandak commander would give the ANA officers a warning order for a Kandak live fire field training exercise (FTX) three days from now. I was tired and went to bed early. The CO said we would be moving to another camp the next day. When I arose the next morning and was packing my things I noticed that my camera was missing. I mentioned it to the CO who told me I could look for it when we stopped at the Kandak HQ at the airport.

Taking the only route the airport I was starting to feel like I had lived here all my life. I knew every bend in the road and where the bumps were as we sped along. I even noticed some of the same people in front of the shops we passed. I waved at all oncoming traffic just like we do in Texas when we're on a country road; some of the Afghans waved back. It was another hot day in northern Afghanistan. When we stopped at the ANA Kandak HQ the CO let me take one of the WFRs back to the Green Howard compound. I went to their CO's tent and met a captain who listened to my tale of woe about my lost camera. He asked me to describe it and when I did he opened his desk drawer and pulled it out. “One of the lads found it last night and turned it in,” he said, “here you go,” handing it to me. I pulled out a ten dollar bill and handed it to him asking that he give it to the ‘lad’ to have a few beers on me. He smiled and assured me that he would. I made my thanks to him, the 1st Battalion, Green Howards, the British Army and all the people of their greatly diminished empire as I backed out of the tent. I returned to the ANA HQ just as the CO came out the door with the Kandak commander.

OOPS! The Lone Ranger Rides Again

The ETT commander and some of his men had gone to town with ‘Doc’ to pick up the SUVs. The CO asked me where the handi-talkies were that we were going to use for communication between our elements when we got on the ground at the Brigade HQ site. I thought the CHOPS was bringing them but he was in the ANA building doing something so I asked the CO if he wanted me to go back to the camp and get them. He looked a little harried so I took his nod for a yes and got back in our WFR and headed back to the ANA camp 20 miles away.

I got around the corner at the entrance to the airport, stopped, took off my IBAS and was ready to go. I was in violation of every rule I knew of for traveling in Afghanistan. I was alone; I wasn’t wearing my IBAS; I had no commo with anyone; and I didn’t care. When I reached the highway I turned right and headed down the road. The WFR was feeling good that morning and it hummed along passing everything we overtook.

I was so familiar with the road now that I used all the time when I had no oncoming traffic to scan the countryside. I was looking for anything interesting I might have missed on previous trips. I saw a camel caravan on the horizon and as I came around a bend I saw another one just a few feet from the side of the road. I slowed down so I could take some video. The man leading the camels stared at me as I held the camera out the passenger side window as I went by. I got some nice footage. The turnoff came into view and I drove past the heavy weapons cantonment area and into the camp. After picking up the radios and grabbing a couple of cold Gatorade's I was back on the road.

Entrance to Qala-i-Jangi, the House of War
(author's photo)

1st ETT CO and author at the House of War
(authors photo)

President Hamid Karzai and 1st ETT Command Group
(Major Cowan's photo)

ANA Brigade Commander with 1st ETT CMD GP
(Major Cowan's photo)

The Corps of Engineers

When I got to the airport I joined the CO on the tarmac on the side of the terminal where he was waiting for the COE people. Things were quiet so I took the WFR and drove down the side of the runway and took pictures of Russian aircraft in various stages of neglect. There was a fairly new MI-8/17 helicopter with "UN" painted on the fuselage and some MIG 21s, a number of twin engine propeller transports and several helicopters, all having seen better days and not going anywhere. I turned around at the end of the runway and saw a helicopter approaching the airport from the south. It was a MI-8/17 with gray paint and no markings.

It landed in front of the terminal shortly and I parked and watched as the CO greeted the COE people. They loaded into the vehicles and we drove through Mazar to the new Corps HQ site. After we passed Qala-i-Jangi there were a lot of shot-up buildings. It must have been a former airbase because there was a Russian MIG-15 on a pedestal at what looked like an entrance to a base. We drove through the building lined street and into the desert. The base looked like it had been wiped off the map sometime in the past. There were some engineer stakes in the ground and a post-hole digger was parked nearby. This was what the engineers had come to see.

The ETT and ANA officers deployed their men into a large perimeter with three or four man outposts every 100 meters. We settled in to wait while the engineers unrolled plans on the hood of one of the SUVs. They would look at the blueprints and then point in one direction or another while talking to each other. I was too far away to hear what they were talking about and decided to read one of the paperbacks I had found at Camp Julian.

As soon as I made myself comfortable, a couple of ANA soldiers came over and hit me up for one of the cold bottles of water we had on ice for our visitors. This was always a problem. For some reason the ANA didn't make the soldiers fill their canteens or they drank them dry early in the day and they didn't have a water resupply plan. I had some bottled water in the back of the WFR and offered them some but they wanted the cold water. I took one of the bottles of warm, or I should say hot water, opened it and took a drink. Then I licked my lips and said, "Gooood."

They watched all this patiently, smiled at my theatrics and pointed to the ice chest so I took another drink and did the same thing but they knew what they wanted and pointed again to the ice chest. I was starting to get worried that I was going to have a water riot when one of the ANA leaders yelled something at them and they took off, one of them with the bottle of hot water that I had offered. Their outpost was nearby and even though I would have liked a cold drink of whatever we had in the ice chest, I made a show of drinking out of the hot water bottle for my ANA friends who were keeping an eye on me.

The day wore on and it was several hours before the engineers were finished with whatever they were doing. The temperature was about 120 degrees and I was on the verge of a heat induced coma when I got a call on the radio. It was the CO telling the ETT to pack it up. The ANA picked up their weapons and headed for their vehicles. The CO got in and we convoyed out of the desert and back onto the street with all the shot up buildings.

What We Are Fighting For

We stopped at one of the buildings and the COE people got out to look at a knocked out BRDM. On the ground around it were a number of Russian Army helmets of the WWII style. They say that the Brits fight for empire, the French for honor and the Americans for souvenirs. There was more war debris and soon most of the Americans were looking around for souvenirs.

I was the only one not interested and climbed a rickety stairway up the side of the building for a better view of the area. This had been a large base at one time and I wished I knew its history. There had been so much fighting in this country for so long that the only ones who knew what had happened here were some of the older residents and men who had been stationed here. There might be some information in Russian but nothing that I knew of had been translated into English. "The Bear Went Over The Mountain"[112] was one of the few sources of information about the Russian experience here during the Jihad but it was mostly about combat operations. I had loaned my copy to an old SOD comrade that I met at the beginning of the deployment and I'm still waiting to get it back.

The Stuff They Left Behind

The souvenir hunt ended and we continued on our way back to the airport so that the engineers could fly back to Kabul in time for drinks before dinner. After waving goodbye to the engineers the CO said goodbye to the Kandak commander and we returned to camp. I was feeling a little restless so I decided to walk over to the heavy weapons cantonment area and have a look. The sentry at the gate returned my smile as I walked by. I passed the volleyball game and continued up the hot dusty road to the tall berm that surrounded the place. There was a sentry there at the only entry point through the berm but he had no objection to my looking around. Inside was a museum of Soviet Military Power.

To my right were several BM-21 multiple rocket launchers. These things were deadly, firing forty 122 mm rockets in volley for more than twenty kilometers. They made a big contribution to the destruction of Kabul when Hekmatyar was fighting with Massoud after the Russians left and the Najib government had collapsed. Each array of launchers was positioned on the back of a solid looking 6X truck. Next to the BM-21s was the artillery. The D-30 was easily recognized from the vehicle recce classes I had with the LRS Leaders Course. There were enough for a battalion. They fired 122 mm shells weighing fifty pounds for 20 kilometers and were towed—a good weapon for Afghanistan.

Anti-aircraft guns were next. Ground mounted ZPU-23-2 towed Triple A pieces were all over the place. They must have been used most recently by the Northern Alliance against the Taliban air force. Mixed in with the ack-ack guns were some multi-barreled 107 mm rocket launchers that could be mounted on a vehicle or the ground. The 107 mm rocket was a mujahideen favorite that could be fired without any launcher; just aim it in the general direction, give it some Tennessee elevation and let'er rip. They were still popular and places like Bagram Air Base were on the receiving end of them occasionally.

On the other side of the yard were the armored vehicles. There were enough T-62 Russian tanks to equip a battalion. Another battalion could be armed with the BMP's that were next to the tanks. The last group of tracked vehicles was BMP's that had been refitted with twin barrel anti-aircraft guns. There were several BTR-60s with anti-aircraft guns, but no BRDMs which littered the countryside. All the armor looked like it was in good condition. There wasn't any ammunition readily apparent which was a comfort because if some bad people got in here with ammo they could make short work of our nearby fort.

The CO walked into the cantonment area, bored like me and interested in looking at the vehicles he had been trained to fight when he was a young soldier. We walked together through the weapons and I quizzed him on the various models just to show off my LRS background. I was impressed that he knew what most of them were and we bantered our knowledge back and forth while I took some more pictures. He was more energetic than me and climbed on one of the tanks and made a Rommel pose before climbing inside for a closer look. I wanted to take a picture of him inside the turret so I climbed onto the very hot metal of a Russian Army tank in Afghanistan and took some pictures. It was hot but that was cool.

After about thirty minutes he had had enough and we walked back to the camp. He told me that he was pleased with the results of our trip so far and now he was looking forward to spending some time with his cousin at tomorrow's CPX. We ate with the ETT that evening and I turned in early having had a long day.

The CPX

The next morning we loaded up and went back to the airport. The ANA had erected several large tents behind their HQ building where the Brigade staff and the two men from MPRI had set up a number of large maps and status boards. CHOPS was involved with this and was all over the tent so rather than add to the confusion I went outside. I visited with the ANA soldiers I recognized from yesterday's attempt to get some cold water. I had come prepared and distributed breakfast bars to whoever wanted one and they were happy.

Inside the tent the Afghan officers were showing their work to the ETT officers and the MPRI men. This was a staff exercise designed to show the progress being made by the ANA to learn the American way of war. The French were conducting officer training at KMTC but after that the U.S. Army was the guiding light.

I dreaded to think that they were actually being required to learn the MDMP[113] or military decision making process along with the risk management process. These two mental exercises were guaranteed to confuse most people with a western education and there was no telling what it was doing to the minds of these Afghans who had been making decisions and living a life of danger for longer than some of their advisors had been alive. As usual, we were training the ANA to conduct conventional combat operations while they were becoming embroiled in an insurgency. It was Deja vu all over again for me as I remembered my Vietnam days and my studies of that war for years afterward.

Like any nation Afghans wanted to have an army for national defense. There didn't seem to be any great threat of invasion by any of their neighbors by conventional means but the Taliban insurgency was obviously being supported by a number of nations around the region and, according to most Afghans, it was being controlled by the intelligence services of Pakistan.

Not everyone is equipped by nature to be a combat leader. Counterinsurgency operations are fast moving and intelligence driven. The MDMP slows things down while attempting to promote logic over instinct or other considerations driven by emotion or ambition. The MDMP gives tools to people who should have chosen another line of work. It also gives them a built-in defense for bad decisions which allows them to advance in rank, hold positions of responsibility, and receive a nice retirement while others clean up the mess.

The risk management process[114] makes cowards of us all. The 'force protection' doctrine by which most of the military was operating in Afghanistan was developed using the risk management process. By evaluating a proposed course of action using the risk management procedures that had been developed by the Army Aviation Safety people, it became difficult to cross the street on an army installation back in the states. A few years later everyone would have to be wearing reflective belts before leaving their quarters.

All of this 'risk aversion' mentality came out of the backwash of Vietnam where we threw away our lives with reckless abandon the same way that young people do to this day. Go down to the auto insurance company and check the rates for drivers between 18 and 24.

Why are they so high? Because young people in that age bracket feel like they're going to live forever and there is no parental supervision.

The Army has done a good job of cutting down on accidents in the field they say, although the suicide rate seems to be increasing. The deaths of off-duty soldiers in that age bracket remains high even after multiple safety briefings that no one listens to. The only thing all of this has accomplished is to criminalize the aggressive nature of the infantry to the point where offensive action will get you in more trouble than refusing to fight.

This was the culture we were trying to infuse into the Afghans. They knew who was paying for everything so they bore these training sessions with good manners but I was pretty sure that they must have considered the last couple of years as one giant WTF moment. 'Insha'Allah' was the way most of them looked at life while trying their best to stay alive another day. Warfighting in Afghanistan was very different than the U.S. way of war and after we leave and take all our technology with us they will return to the old ways.

Their enemies on the other hand, have evolved in the way they fight. Along with not appearing to fear death in battle they have developed 'smart bombs' that utilize the human brain wrapped up in explosives and wound-producing shrapnel like nuts and bolts. They can put one of these 'smart bombs' right inside the office of a high official and blow him to bits without putting a hole in the roof. Between the suicide bombers, the improvised explosive devices (IED), and the vehicle borne improvised explosive devices (VBIED) and whatever other methods they were using to kill the infidels who had invaded Dar al Islam, 'Insha'Allah' they would kill enough of us, regardless of our MDMP and Risk Management, to cause the people of the U.S. to pull the plug through their elected officials and leave.

Durdee Time

I pondered these things while I waited for the CPX to conclude. Finally the ANA officers came out of the tent and headed to the HQ building for lunch. The CO came out and told me that we were invited and he wanted me to attend. After lunch we would show a DVD I had made back at Camp Julian with the video footage I had taken of the Victory over the Russians parade at the end of April.

I followed him into the HQ and entered a large room where we would eat. There was a large plastic sheet on the floor with food and drink. We all sat on the floor and were served dishes of rice heaped up over a piece of gristly meat and of course, the durdee.[115] There was a drink being served that looked like milk which I did not want because I was lactose intolerant having stopped drinking milk as soon as beer became available so many years before. I wasn't drinking beer anymore but I wasn't drinking milk either.

This was one of those 'hearts and minds' moments that my Special Ops mentor, Bil had prepared me for years before in the drug war. I saw several ETT members decline when the pitcher was offered to them and when it was my turn, I held my glass out so that the server could fill it. The ANA officers were all watching this with interest.

I realized that I needed to drink some now. The thought of projectile vomiting and gut wrenching diarrhea was weighing heavily in my risk management process of considering any food or drink not provided by the U.S.A. or its western allies. I took a small drink. It tasted good. It was foamy and tasted like cucumbers with some liquid yogurt and was lightly salted. I took another drink, the eyes of the ANA officers following my every move. I kept it up until I emptied my glass. There were smiles all around on the faces of the ANA; I had passed a test. I actually liked the taste of the drink so much that when the server had finished giving everyone a glass I held up my glass and asked for some more. One of the ANA officers laughed and directed the server to give me another.

The attention shifted away from me and the conversation around the table increased in volume as the ANA officers conversed among themselves and with their ETT counterparts through the interpreters. I continued to work on the huge plate of rice in front of me while grabbing a few pieces of vegetable sticks from plates on the table. I was really full and had a lot of rice to go. I looked around and saw that the rest of the people were in the same boat so I continued to nibble while sipping the delicious drink. Finally, the ANA Brigade commander said something and the meal was over. The plates were collected and then the plastic sheet was picked up and taken out leaving us sitting on the floor and ready for the next event. The Brigade commander made a short speech thanking the ETT and MPRI for our assistance in the CPX. Several other officers and our CO said a few words of thanks and then it was time to show the video.

The Big Parade

The Victory Parade was a major pain for all involved. Preparations for it had almost stopped any other work at KMTC and the time was now known as 'marching season.' The ANA officers had been through the same ordeal, marching for several hours each day, in preparation for the one time their soldiering skills would be scrutinized by the civilian authorities. The military parade has been used as a measuring stick by military and civilian leaders for this purpose for thousands of years; the logic being that if the leaders can force the soldiers to march in step and be aligned with the person next to them then the leaders are doing their jobs and the soldiers must be prepared for war.

No soldier that I know ever subscribed to this logic and all of them hated dismounted drill. The only parade that any of them ever wanted to march in was a victory parade for a war he had fought in. Even then, when the Desert Storm soldiers marched in the big parade in New York City after the war I heard it was a giant pain and everyone there would have rather watched someone else march on TV while having a 'Victory' beer with their comrades. There hadn't been any 'Victory' parades for the Korean and Vietnam Wars and it didn't look like there would be one for the drug war or the wars of Iraq II and Afghanistan.

Afghan Problem Solving

The ANA had set up a small TV and a DVD player. I handed my DVD to an ANA captain and sat back. He inserted it and pushed 'play' on the remote. The first scene came on and then the image froze. The captain pushed play several times but the image remained frozen on the screen. He ejected it and tried again. The video advanced a few frames and froze again. He looked at me. I suggested that we eject the DVD again and unplug the DVD player then try it again. He did and we got the same result. I didn't know what to do. The DVD had worked perfectly in my laptop and there was no reason for it not to work now. I told the captain that I didn't know what else to do and that's when he showed me a trick.

He ejected the DVD again and held it with one hand. With his other hand he drew a hunting knife from a sheath on this belt. He took the knife blade and began to cut around the sides of the hole in the center of the DVD. He drew the knife blade around the hole several times. Then he wiped away any debris and put it back in the DVD player. "Right," I thought, "like that's going to do any good." These Afghans had been living in the Stone Age before we got here and now one of them was going to carve a larger hole in the DVD and that would make it work.

That's exactly what it did. The video came on and played all the way through without a stutter. The ANA and the rest of us were amazed and the captain's solution to the problem so impressed everyone that we had to watch the DVD a second time after we finished congratulating him. The Afghans had learned to improvise as a way of life while their country was destroyed by invaders and in-fighting and this was but one example of the ingenuity of these people.

The ANA officers watched the video and commented now and then which brought laughs from the others. When their unit came into view there was some more laughter. They looked pretty good but their standard for lock step marching in the Russian style was high so they would point out any errors they saw. They enjoyed pointing out the errors of the other units just like any other soldiers I have known so some things were the same in any army while other things were worlds apart.

When the video finished the brigade commander dismissed his staff and we took our leave. I gave the CO a copy of the DVD to give to the brigade commander and I gave one to the captain who had saved my face. I gave another one to the ETT commander to give to the Kandak commander and told him that if they wanted more I would make copies at the camp and give them to him tomorrow. We drove back to the camp and I waited for the drink to make me sick but it never did. I went to sleep early that night and didn't wake up until dawn.

The Blue Mosque

The next morning we drove back to the airport and joined the Kandak commander and several pickup loads of ANA and headed out. We drove into Mazar and stopped at the Blue Mosque. The Kandak commander wanted to show us around the most famous place in northern Afghanistan. We walked onto the grounds and the Afghans who were hanging out there took a somewhat hostile interest in the activities of us ferengis or infidels who were walking on their religious site. The ANA commander didn't seem to be concerned but I noticed that everyone nearby was taking an intense interest in what we were doing.

I called the CO's attention to the situation and he noticed immediately that the atmosphere around the mosque had gotten tense. There were a lot more Afghans in the area than there were ANA and ETT soldiers and after being reminded of what had happened at Qala-i-Jangi I mentioned to the CO that maybe we should not go inside and just take a few pictures outside and leave. He agreed and told the Kandak commander his decision. We took pictures with every Afghan eye following us until we got in our trucks and left.

Mazar-i-Sharif-West Camp

The next leg of our journey took us west through Mazar on the main road. We drove through another gate in a large wall like the one at Qala-i-Jangi and out another after crossing a large area surrounded by walls. Then we were on the open road. As we sped down the highway, I saw a large pipeline on our left, a natural gas pipe from Sheberghan to a power plant and a fertilizer plant in Mazar. Beyond the pipeline were open fields with an occasional cluster of large bushes. The CO said that the last time he was here the fields were covered with opium poppies. "Do you see the large green bushes out there in the fields? He asked. "Those are marijuana. When the workers get tired of working the poppies they take a break in the shade of the marijuana."

The CO then began a long explanation about how we needed to stop the opium production before the western world found out and cut off aid to Afghanistan. I wasn't too concerned about that having worked the drug war in Texas for the past fifteen years; no one was doing anything drastic to Columbia or Mexico. The British had been given the mandate for this piece of the Afghan puzzle and we would have to wait and see how they did. The biggest problem was the local population that had been growing poppies for a living for a long time. The latest great idea was to get the opium growers to plant saffron, a flower that produced an expensive spice. Time would tell whether the Afghans would buy into this scheme.

We were approaching the ANA camp. There was a lot more vegetation on the west side of Mazar. The camp was a short distance from the highway but concealed by some walled compounds and vegetation. We left the highway and drove up the road until we saw the HESCO walls of the camp to our right. The sentries shouted as we approached the entrance, then saluted the Kandak commander and waved us in.

This camp was about the same size as the other and laid out in the same manner. The mortar pits were well made and the rest of the camp had a neat appearance, meaning that they had either done a lot of work that morning or they kept the place neat all the time. They were squared away. I got out and walked around a bit. The ANA soldiers were friendly and I watched some of them play volleyball on a court on the other side of the wall.

Just beyond the volleyball court was another heavy weapons cantonment area. After we ate lunch, while the others had a siesta, I walked out the gate and over to it. It was an LRS vehicle recognition examination. Artillery pieces of every caliber from smaller anti-tank guns to 152mm towed howitzers were lined up in rows. T-54 and T-62 tanks had their own rows. I wandered through taking video of the equipment until my battery went dead.

A Small World

After I returned to the camp I thought about going back with my digital camera but since everyone else was crashed out I went down to the medic tent. LTC O, the PA who served as the ETT doctor, wore an SF tab and jump wings. He was sitting down reading something when I walked in and he invited me to sit down and have a Gatorade. I took him up on his offer and we began to get acquainted. Somehow we got to taking about units we had served in and I told him about G Company when it was one of the last Airborne Ranger Long Range Reconnaissance Patrol companies in the army.

I mentioned Gordon Rottman, the company's operations sergeant and well known author of military history books. He chimed in with, "I went through training group with him." He then told me about going through the SF Q course back in 1968. He missed going to Vietnam, had left SF and gone into the medical field. He was a practicing PA when the war started. He decided that he didn't want to miss this one so he volunteered to go to Afghanistan when the Oklahoma National Guard was alerted for this mission. He had managed to arrange for a transfer to an SF unit when the Oklahoma unit went home; he liked it in Afghanistan.

He was into haircuts though and prodded me to get one every time he saw me but didn't let my lack of interest in high and tights interfere with our getting along. He was a LTC in the Medical Service Corps and I was a Master Sergeant of Infantry who went to Vietnam so I hoped he would understand how I could nod and mutter words of compliance when he was ragging on my haircut and then not do anything about it.

Kandak Live Fire

That night after dinner we had an informal briefing on tomorrow's activities. We would drive back across Mazar to a training area about half way between the airport and the eastern camp. There the Kandak would conduct a live fire exercise utilizing their mortars and SPG-9 recoilless rifles in support. There would be fire and maneuver and a final assault on the objective. The ETT had been training with them for the last eight months and this would be a kind of validation exercise for all of them.

The next morning before dawn, the ANA loaded into trucks and we followed them as the Kandak commander and ETT commander led us through Mazar and onto the training area. It was about 0900 when we arrived at the site. The other half of the Kandak was already there and the newly arrived soldiers moved quickly to the areas designated by their officers as the assembly area. The other mortar and SPG-9 crews moved their weapons to the positions already occupied by the heavy weapons elements already in place.

The mortars were set up in a low area behind the assembly area which was concealed from the objective by a low ridge line in front of us. The SPG-9s were set up on some high ground to our right rear where they had line of sight with the objective. The soldiers were issued ammunition and the officers began moving their troops to release points just behind the low ridge to our front. I positioned myself where I could get a good shot of the mortars because I knew that they would initiate the attack with a barrage. I watched them getting ready to fire; the ETT guys were scurrying around making sure that everything was good to go.

I heard some commands and started my video camera. There were three mortar tubes set up and each one had a man hanging a mortar round and looking at the one giving the commands. Then one after another, they dropped their rounds down the tubes and the exercise was on. Each round flew out and into the air. I turned around and aimed my camera on the objective and soon the three rounds detonated within the area designated as the objective.

Then the SPG-9s fired and those missiles sped across the valley and struck some walls of what were ruins of old military buildings. The mortars fired again and then the PKM machine guns of the ground troops chimed in. Several RPG gunners added their projectiles to the clouds of dust caused by the explosions on the objective. The fire on the objective was sufficiently intense for the maneuver element to be released. The rifle platoons left their release points and got on line in trenches just forward of the low ridge.

I moved up to a position just behind them and videoed the advance of the infantry. The PKMs were firing to the right of the maneuvering infantry and RPGs were landing in the area of the objective. Now the infantry was firing their AK-47s as they bounded across the open area toward the objective.

There was no cover or concealment for the infantry after they left the trenches. They had to cross several hundred meters of billiard table open ground to get to the objective. If there was a determined enemy on the objective who decided to stand and fight they would have been decimated. That wasn't the problem though. In the burgeoning insurgency that was on the horizon, finding and fixing the enemy in large groups and maneuvering a battalion size element to finish them was the least likely scenario that they would ever face.

This was a typical training exercise devised by the ETT to mirror similar exercises that they had performed during their own training back in the states before they went to war. It was a minimal risk exercise they could be expected to perform with no training casualties but it did little to prepare them for actual combat and there were few exercises that could. At least they had learned how to use their indirect fire weapons in concert with their supporting direct fire weapons and that could help them when they did go into combat.

The line of infantry now moved forward firing from the shoulder as they marched, right out of the 1955 infantry manual that we used in the 158th Infantry at Fort Huachuca back in 1963. The thought was that now that they had achieved fire superiority, they would conduct a final assault and sweep across the objective. This they did and after about ten minutes, 'cease fire' was called and that was it. The whole thing had taken about 30 minutes and now it was over. I saw the CO and the Brigade commander watching from another vantage point and they seemed to be pleased. No one was hurt and all the components of the Kandak had worked together, all in all a successful operation.

We loaded everyone up in the trucks and the west side element of the Kandak returned to their camp. We drove ahead of the eastside element and returned to our camp which was nearby. The CO was happy about the events of the past few days and told me that we would be returning to Kabul the next day. There had been some talk about a visit to the ETT element in Kunduz, the next large city to the east but now he explained we would return to Kabul by the RAF C-130 that had brought us here. I was a little disappointed because I was enjoying our trip and wanted to stay in the field for as long as possible. It was kicking my butt but it was where I belonged.

Back To Kabul With The RAF

We packed everything into our WFRs and said goodbye to the ETT guys. They would be coming back to Kabul in the middle of August and I planned to give them a copy of the video I took of them during the FTX. We drove out of the camp and drove back to the airport one more time. I recognized several of General Atta's militia at the big checkpoint and waved to them and they waved back. It was like we were all working toward the same goal and we all knew each other; I loved Afghanistan.

We drove into the airport and stopped at the ANA HQ where the CO said goodbye to the brigade commander. I saw the captain who had saved my face and after saluting him gave him a hug in the Afghan way and rubbed cheeks. I wished I could stay with these guys and do something to help them rebuild their army and their country but it was time to go. The CO's cousin and his Ranger buddy joined us and we drove out on the tarmac to wait for the RAF.

I talked with the others and took a few pictures until one of the younger soldiers heard a plane and started scanning the sky. Soon someone spotted the C-130 coming in low on its final approach. It slid in and reversed its props almost immediately making a loud noise for a few seconds and then taxiing off the runway and onto the tarmac where it turned around and lowered its ramp. The engines were running as the loadmaster and his helpers drove our WFRs onto the plane and tied them down. We filed onboard and took a seat and waited. Without any fanfare the plane lurched forward and began taxiing down the runway. The pilot turned it around again and gave it full power and we rolled down the runway until it leaped into the sky and headed for Kabul.

The Hunt For Bin Laden And The CIB

At the Kabul Airport we said farewell to the CO's cousin and his Ranger buddy and returned to Camp Julian. First Sergeant O was running the TOC and things were a little more squared away. I joined the roster for TOC duty where I watched and soon learned how to play spider solitaire so that I could add my psychic energy to the hunt for bin Laden. I kept my ear to the ground hoping to get down to Gardez but everyone else wanted to go there to try and get a CIB.

Down in Gardez, the ETT had been rocketed and there had been at least one firefight involving its members. They were pretty excited about the prospects for badges and medals according to one of their members who had been sent back to the HQ for reasons no one was talking about. The badge hunt was on everywhere. I thought that Colonel Knight's quest was an aberration but if he had a disease it had reached epidemic proportions in the 45th Brigade.

The ETT CSM's son, a member of the security force at Camp Phoenix, was involved in a supposed shooting incident in the village across the street from Camp Phoenix. No bodies or weapons were found but the patrol claimed that they had been under fire and their case was under consideration by the CG for CIBs.

Trigger Time

There had been a shooting incident out in Pul-e-Charkhi where a 45th soldier opened fire on a couple of Afghans who he said were about to engage him with an RPG. Fortunately for them he missed. A search of the area yielded a piece of broken pipe that the Afghans were attempting to salvage from a bombed out structure. After an appeal was made to the troops to be sure before shooting, the whole incident was forgotten.

In another shooting incident, the CG's personal security detachment had lit up a vehicle which they said was following too closely behind them on a trip through Kabul. It's amazing that more Afghans weren't shot this way because tailgating was how they drove in Afghanistan. Everyone was in a hurry and the vehicle in front was a challenge to the one behind it. The security guys put a number of rounds through the radiator of the car and the Afghan was lucky that they didn't put them through his windshield. The CG wasn't impressed and read the young soldiers the riot act, reminding them of the 'hearts and minds' concept that seemed to be so easily forgotten.

The Modern War Video

A few days after we returned I got a notice that some large packages for me had arrived at the post office at Camp Phoenix. I caught a ride that afternoon on the BUB run and picked them up. It was the editing machine and monitor I had asked for. They had been shipped halfway around the world for about $30; this was truly a 'modern war.'

I set up the monitor by suspending it from the bottom of the top bunk where I stored my equipment so I could lay on my back and edit. The first thing I did was load several hours of tape and then I made rough cuts of each part that I wanted to use. I decided to show a brief history of our pre-deployment training at Fort Carson because I knew that the 45th Brigade had been trained there as well before they deployed. Then I showed scenes of basic training at KMTC where the Afghan soldiers that the ETT advised were trained before they became an operational Kandak.

A major had shot some video while he was in Gardez and he offered it to me. He had used a Canon 3 chip camcorder on a tripod; the video was beautiful. The ETT in Gardez had driven there in the wintertime. The mountains were covered with snow as was the FOB. He had great footage of a live fire FTX exactly like the one we had just observed in Mazar. There were some excellent scenes taken from the cockpit of an MI-8/17 which gave a great overview of the countryside around the FOB. I used a great deal of his footage to tell the story of the 1st ETT. Finally, I added my own footage of the trip to Mazar and left all the edited video on the machine.

I wanted to wait until we did the RIP with the next unit before I completed it and gave them copies. The CO checked on my progress every now and then and told me that he had some great music that he wanted me to use. I got a copy of the tape and began to listen to it so I could try and match the music to the motion. It was a great distraction because life in the camp was pretty boring after the trip to Mazar.

My Friends

One day I got an email from one of the TAG guys letting me know that a delegation of interpreters from KMTC wanted to come to Camp Julian and bring me a gift. I answered and asked him to tell them my schedule and how to contact me when they got to the camp. The next Friday, JS and a number of the interpreters came to the gate at the camp and the Canadian guards called our TOC. I got the message and ran to the gate to greet them. I wished I could have brought them all into the camp but it would have taken a long time to clear them so we decided to have our visit outside the camp. They had a lot of bags so we walked over to the bazaar area and sat together under a little shelter.

JS made the formal presentation of a Pakol or 'Massoud' hat and a Karakul or 'Karzai' hat. They gave me several Afghan scarves and by then I was getting embarrassed. I knew these men didn't make much money and they had spent a lot for these gifts but they were just getting started. Next they gave me a traditional pair of Afghan pants, baggy legged and a huge waist with a draw string. With the pants came a finely woven shirt with small decorative white beads on the chest and neck area.

I was starting to get overwhelmed by the generosity of these men but they weren't finished and one of them pulled out a beautiful green dress with much gold embroidery for my wife, with it came a florescent green veil. Finally they gave me a vest for my son with gold embroidery. I was torn with emotion. I wished I had some money to give them but that would have been an insult. I wanted to tell them how much I had come to love them but that seemed superficial. I don't know if I did anything to advance our cause in Afghanistan while supervising the interpreters at KMTC but I felt like I had touched their hearts and they had touched mine. We took a few pictures and then they left.

A Blast From The Past

I made quite a few trips to OMC-Alpha as an extra gun. On one of them I ran into Colonel Knight who was now stationed there. After what I presumed to be his unsuccessful coup attempt, he was relieved and disappeared from KMTC and Camp Phoenix. I had seen him one other time before I was relieved and 'disappeared,' and that's when he dropped by the Alamo and came into our new mess hall with another LTC for lunch. Apparently his presence was made known to the CO who came storming in there a few minutes after his arrival and demanded to know what Colonel Knight was doing there.

The scene was so tense that I moved to a better vantage point where I could tackle Colonel Knight if he made any aggressive moves. After my conversation with him over the game of pool I considered him to be the malefactor in this 'officer business.' I was prepared to assist the CO despite my own antipathy towards his command style. The CO demanded that Knight depart immediately and he refused. Then the CO ordered him to come to his office immediately and he said, "Only if I can bring a witness," nodding at his companion. The CO must have agreed by some non-verbal communication because they all strode out of there like they were on their way to an important meeting. I looked at the CO and offered myself as a backup as he walked passed me; but he waved his hand for me to remain seated. That was the last time I saw Colonel Knight until now.

He bore no apparent animosity toward me. After asking me what I was doing there and finding out that I had been transferred to the ETT, he laughed and said, "I'm surprised that you lasted as long as you did." He was happy at OMC-Alpha, he said, because the 'social opportunities' were much better there. He had once complained to me about the lack of opportunity at Camp Phoenix for female companionship and when I told him that it was for good order and discipline he threw that back in my face by reminding me that we didn't have any such restrictions in Vietnam. I countered with the trump card, "We lost that war, sir," which ended that conversation.

He had been involved in a wreck with our female supply clerk shortly before he disappeared. They were in a single vehicle convoy-one vehicle roll-over out on the range near KMTC and there was some speculation that connected that incident with his disappearance but of course since I was the Intelligence Officer for the TAG, I knew nothing.

Fobbit HQ

OMC-Alpha was situated near the new American Embassy. This was going to be the largest U.S. Embassy in the world. Behind high walls, several cranes were working on multi-story buildings. The people at OMC-Alpha were in their own world. Their status would soon fall under the classification of 'Fobbit,' the new term for the people we called 'REMFs in Vietnam and afterword. 'Fobbit' was a play on the acronym FOB or forward operating base.

A 'hobbit,' on the other hand was a mythical creature from the imagination of an English author who had written a series of books about wars and quests which captured the attention of my generation and were later made into a series of popular movies.[116] The hobbit was a creature that lived in a small village known as a shire and never left it because of the dangers that existed elsewhere; hence the 'fobbit' who never left the 'wire' of the FOB for the same reason.

Going 'outside the wire' was still the badge of honor for the troops who went out. To the fobbits, who were now the majority of the armed forces, it was the sign of a less intelligent service member who did those kind of things The 'Revolution in Military Affairs' that was being written about in all the service journals and commented on by this generation's military analysts, was much more technology driven than tactically innovative. Robotics was the wave of the future and the shooters watched their mission slowly devolve into guard duty and daylight patrols while the killing was done from above.

Air strikes and drone strikes were killing most of the ones who opposed our presence in Afghanistan. These required large logistical tails and few if any actual combatants. There were some drone operators who were awarded Bronze Stars for flying drones by satellite telemetry in Afghanistan from air conditioned offices in the U.S. These were rescinded after howls of anger by the 'shooters' and replaced with other medals for meritorious service. As I write this, there has been another medal that was approved by one SECDEF and withdrawn by the next that was made for these warriors who fight the war with drones and others who attack enemy computer systems on the internet.

In 2005 the U.S. Army established the CAB or combat action badge. This was supposed to recognize a non-infantry soldier's service in actual combat. There are regulations for awarding these badges. The CAB joined the CIB and CMB[117] as prizes for the hunt and the rush is on to get one of these 'see, I been there' indicators for all to see when the soldier returned from the war. As in other wars, most of the ones who really earned the badges considered it with painful self-deprecating humor when remembering their days of youthful folly when they considered these things worth risking their life for.

President Hamid Karzai-The Parade
(Major Cowan photo) .

And when you hear of wars and rumors of wars, do not be alarmed; it must happen, but the end is not yet.
Mark 13:7

Chapter Ten: August-September 2004

No Joy At Camp Julian

Meanwhile, back at Camp Julian the fight continued. The ETT was on the home stretch and rumors were flying about when the unit would actually leave. The war in Iraq had screwed up a number of deployment time schedules and the Oklahoma boys were more than a little concerned about how long they were actually going to have to stay there. The Specialist that had incurred my wrath the first day I had arrived slept two bunks away from me and spent most of his waking hours complaining about everything under the sun. He was so distraught about the situation that I began to fear for his safety.

Part of me would have been happy if he had shut himself up with a bullet but the stronger part demanded that I do something. The CSM lived at the other end of our building and was fully aware of the Specialist's condition. He had somewhat resigned from any active participation in the daily activities of the ETT along with the operations officer and a few others who were there in body but their minds seem to have gone somewhere else. I went to him repeatedly and expressed my concern for the mental health of this specialist and the possible consequences of ignoring him like they had been doing for some time. The CSM appeared to rally in this case and began spending a little time with the specialist which seemed to have a calming effect. Then they sent the specialist down to Gardez for a couple of weeks.

That just about sent me around the bend. Here I had been angling for a trip to the combat zone where I felt that my years of service had prepared me to strike a blow to our enemies and they sent this idiot instead of me. There were a few others like me that wanted to go there and we all commiserated about the general unfairness of army life. Oh well, at least it was quiet in the barracks for a while.

The Arg

I started going to the palace with the CO. The ANA brigade commander was stationed there and they would meet outdoors under some large shade trees in a courtyard near the King's palace. President Karzai lived in the same complex and we passed the entrance to his compound every time we visited. The grounds were very picturesque and I shot a lot of video of the CO with the ANA brigade commander.

Our CO had done a good job; his skills as a politician helped but his character had done more to gain the respect of the Afghans. The CO and the ANA colonel had become friends. Now they just made small talk because the CO was leaving and our replacements would soon take over. The whole process of building rapport and trust would begin again. It was tough for them to face the fact that although they had invested a lot of emotion and sweat in the performance of their duties, their friendship was coming to an end.

The ANA colonel made it known that he was worried about his future because the Afghan government had very little infrastructure that was operating at all and things like job security and retirement plans were things to wonder about, not enjoy. The Afghans envied our system. They hoped that they too would eventually have similar benefits like we enjoyed but for now they could only dream and try to get something for themselves while we were here. The corruption that everyone complains about was no worse than it is in many other places. The culture, religion, and the wars had pretty much destroyed the concept of 'future' that we put so much stock in. Most Afghans didn't dare ignore a short term benefit today in exchange for a more substantial benefit tomorrow.

The Kandak at the Arg was literally the 'Palace Guard.' Of course there were several rings of security around President Karzai, and the ANA soldiers made up a small part of his security. Most of the security close to him was provided by contractors, former SOF who were cashing in on the security business gold rush produced by this war. Being the president of Afghanistan was a dangerous job and those guys earned their pay because he is still alive after ten years in the saddle.

The Final Cut

I showed the CO a rough cut of the video and he liked it until we got to the live fire FTX in Gardez. I had inserted a scene where the ETT guys were making a target for the SPG-9s. The men took a 4'x8' piece of plywood and spray-painted the outline of an SUV with the name of the 45th Brigade's CG above the silhouette. The next scene showed the Afghans making a direct hit on the plywood with their SPG-9, blowing it high into the air with the explosion of the projectile. The CO didn't think that this would enhance his reputation with the CG and probably felt that it could negatively affect his OER so he asked me to take it out. I pressed him to keep it in, 'cinema verite' and all but I could tell by the way he looked directly into my eyes when he asked me again to take it out that he meant it, and I did.

Standing Operating Procedures

I had made peace with the commo officer. While we were in Mazar, I gave him the silver button with the Hammer and Sickle that I had found. We continued to work on a commo SOP; I wrote it and he approved it. The ETT didn't have one that covered the current commo equipment and was based on their experiences in Afghanistan. I had started one that covered the Harris HF radio. Now I worked on TOC procedures and the commo procedures between the TOC and the vehicles driving around Kabul.

We kept track of every vehicle that left Camp Julian, including the departure time, destination, route, ETA, and actual arrival time. I got a vehicle designated as the QRF vehicle and kept it parked near the TOC in case we had another vehicle breakdown somewhere in the city. We now had tow-straps in the back of every WFR assigned to the ETT. First Sergeant O and I worked out the procedure to be followed if the need for a quick response arose.

All of this was iffy because the radios wouldn't work in certain parts of the city and the guys didn't follow the procedure every time they went out. The CO and some of his staff went to the palace almost every day and the routine had set in. We pushed the CSM and the CO until they bought in to our recommendations and we started to feel like we were doing all that could be reasonably expected of us under the circumstances. In the end all the SOPs and planning couldn't replace the good judgment of the man on the ground and that's what we were all counting on to get us home alive.

The Long AT Comes To The End

The ETT guys had found a few restaurants in Kabul that they went to from time to time. This was strictly prohibited by the rules but many of the 45th Brigade officers and senior NCOs were doing it. There was a place run by a women from Thailand that the guys liked and there was a pizza place not too far from Camp Julian that was also popular. I was game for anything that got me out of the camp so I went to these places as an extra gun whenever I was invited.

We were in the last stretch of the deployment and it was starting to look like a National Guard AT coming to an end. It reminded me of summer camp at Fort Hood when I was an intel weenie with the G2 section. During the second week of AT we would all go to the field but by Wednesday many of the senior people had gotten motel rooms in Gatesville, a small town near North Fort Hood, and they would ride out the last couple of days in town.

The ETT element from Kunduz came in. They had been replaced and were now the first team out of the field. They all had an attitude, being 'real' soldiers and not 'fobbits' like those of us who worked in the TOC. I knew how they felt and left them alone. They stayed to themselves and when we saw them, they were either going to chow or hanging out in a dayroom near our area where people watched TV and read books.

Allies

I saw one of the senior NCOs getting very friendly with a female Canadian soldier and considered saying something but by now so many of the rules were being ignored by the troops it was a matter of enforcing the ones that were most closely connected to safety. Confrontations were avoided by most; when I got into it with some of these guys who tried to take a vehicle without following procedures they resented me as someone they didn't know who was not respecting their status as combat veterans. This would set the tone around the HQ until we left.

A Belgium airborne unit had moved into the camp. They had the mission of pulling security at the Queen's Palace that overlooked Camp Julian. They were some big boys and they started using our shower and latrine area. The vibes were getting bad. There was a lot of complaining but in the interest of coalition warfare nothing was done and everyone adjusted.

Some of our guys got tired of standing in line for a shower in our area and started using the shower designated for the civilian workers from Nepal who worked in the kitchens and did the laundry. The Nepalese usually showered with their underpants on and when one of the Oklahoma guys walked into the common shower room completely naked the Nepalese fled and one of them complained. These complaints eventually got back to our HQ and the CSM tried to get everything straightened out but this would go on until we left.

Corruption?

There were some Conex containers that the ETT used to store extra equipment. One day I was asked to help with an inventory of one of them. When the door was unlocked and opened I got some insight into the problems we faced in trying to rebuild Afghanistan. There were hundreds of new blankets from Spain, apparently donated by the Spanish government, each one sealed in a zippered clear plastic bag. Next to them were stacks of portable radios with hand cranked generators for recharging the batteries.

These items had been given to the ETT for distribution to the Kandaks they were supporting. I don't know why there were still so many blankets and radios left or if they had ever been distributed in the first place but now we had a boatload of blankets and radios to get rid of. Someone made the decision to give the blankets to the ETT and I got a nice one with a large red rose. Before the radios disappeared, I got one for Matt over at Camp Phoenix.

Two years before, President Bush had called for a new 'Marshall Plan' for the rebuilding of Afghanistan. Since then, the U.S. had invaded Iraq and Afghanistan was becoming the forgotten war. The vaporization of the Taliban and Al Qaeda foot soldiers in 2001 had created the sense that the war was over. Those who knew the history of Afghanistan weren't so optimistic and the lull in the fighting during 2004 wouldn't last.

Another Team Returns

The ETT from Mazar returned and the whole cycle of jealousy and hate repeated itself when they found the restrictions of life at the flag pole intolerable for the 'combat' veterans that they were. I continued to work on the SOP that we would present to the unit relieving us so that they could avoid some of the problems that we had to deal with; I also knew that we might as well have written it on toilet paper where it would have gotten some real use but it was hard to get our printers to use it.

More SOP's

The deputy CO was pleased that we were working on the SOP. He began to review our work every day and suggest other functions that could be described in detail. Since there wasn't much else to do during TOC duty other than to continue the hunt for bin Laden in the various versions of solitaire that Mr. Gates had thoughtfully left on his software solutions, we added new annexes, appendices, and tabs daily. Of course, when you present an SOP to someone you have to do so with graphic displays that have a twofold purpose.

The first and most important is to demonstrate your mastery of Power Point, the briefer's essential companion; its skillful use had become an art form in staff circles and without some flashy PPT slides you might as well stay in your seat. The second is to try and winnow down a lot of information into a few 'bullet' points that would remain in the minds of those being briefed. After you have sat through as many briefings as I had to during my first six month in Afghanistan you begin to wish for a real bullet to either shoot the briefer or yourself depending on your particular personality. I was a survivor so I had to call on the Lord a lot as his word in 2nd Corinthians chapter ten reminds us that our battle is not against flesh and blood.

TAG

My old unit, the TAG was getting ready to leave. I wanted to stay in Afghanistan as long as possible so I told the CO that if they asked for me to return to Camp Phoenix to go with them, I would rather go home with my new found friends. He said he would take care of it and

I never heard a word about it when they left. After they were gone I made a couple of trips back to KMTC to visit some people that I wanted to say goodbye to before I left. One of them was a high ranking staff officer who I thought a lot of. I gave him a copy of a book of Afghan military history and he gave me a copy of a history of Afghanistan that was printed in English.

I didn't go to the Alamo and limited my visit to the Afghan army officers I worked with and of course one more visit to the interpreters. Most of them were working but JS was there and I gave him the letters of commendation that I had written for each interpreter. The basic letter was the same but I had tried to write some individual comments about each man's work. At the end of all the letters I recommended that these men be allowed to immigrate to the U.S. if they wanted to. How was I to know that in 2007 the U.S. would begin to offer special visas for interpreters who had worked for a year or more for the U.S. forces. That was the least we could do seeing that working as an interpreter for the U.S. was a death sentence as far as the Taliban were concerned. We hugged each other one more time and said goodbye.

All Present Or Accounted For

The 'Cowboy' team from Gardez was the last of the 1st ETT to return to the fold. They pulled in with a Humvee that had been hit by an IED a few days before. Some of the front fender was gone but it was still drivable. Apparently the Taliban guy had gotten buck fever and pushed the button on his cell phone a split second early and this ensured that the ETT would return to Oklahoma with all its personnel alive. They were the most sullen of the teams, actually having been in some combat.

I wasn't very sympathetic to all their antics but I tried to show that I was available if someone wanted to talk. I knew that some of them were suffering from some event that had taken place. They were keeping to themselves and not even mixing with the ones they knew on the other teams. During my Vietnam combat time we had lost almost our whole platoon in a couple of hours and I thought that these guys should be praising the Lord for all he had done for them; but they continued to keep to themselves and the lid would stay on until we got to Fort Carson.

Awards and Decorations

One of the main activities now was the paperwork required for the awarding of medals. Each award required the filling out of a long form with lots of information about the individual. Citations regarding the actual activities of the individual that merited an award were needed and in some cases witness statements were needed. Each team had a personnel officer who stayed pretty busy in the days before the CG would come and hand out the awards.

When we arrived at Camp Phoenix back in December, the 45th Brigade CSM told us that we wouldn't get a medal at the end of our tour if we had a negligent discharge while clearing our weapons at the clearing barrel at the entrance to the camp. Since that was the only criteria for not getting a medal, I was curious to see what kind of medals everyone would be getting. There was quite a bit of excitement about the subject and I could see that the Oklahoma boys were all hoping to go home as war heroes. When we left Fort Carson they told us that we were all heroes and I passed that word on to several of the soldiers who were discussing the upcoming awards ceremony. They didn't appreciate my gentle sarcasm and didn't bring up the subject again in my presence but it was the hot topic around the camp.

The awards ceremony day was scheduled for the 1st ETT. The CG would come and award campaign medals known as the Global War on Terrorism Expeditionary Medal and individual awards to those who had been selected for exemplary service. I took pictures of the ceremony and First Sergeant O pulled duty in the TOC because neither of us was being honored that day. The campaign medal was automatic and I would get mine at Fort Carson or in the mail. This was a fitting end to my own quest for glory. I had not received any medals when I left Vietnam, much to my disappointment and now I was not receiving any, much to my relief.

Since there weren't any valor citations being awarded at this time, they were still being worked on at the team level, the highest award being given that day was the Bronze Star for Meritorious Service. This was the medal that seemed to be the most desired. Generally speaking, it was the award given to field grade officers at the end of their tour and only a few enlisted were to receive one.

One sergeant major that had been in the Humvee that was damaged by an IED was given a Purple Heart for injuries sustained in combat. He didn't have any outward signs of wounds so I asked someone on his team what had happened and was told that the sergeant major had major damage to his ears, injuries that qualified him for a Purple Heart. He also received a Bronze Star making him the most highly decorated member of the 1st ETT. I didn't know what any of these men had done to deserve these awards while deployed so watching the awards ceremony caused me to remember similar occasions in the past where some of the men receiving awards were held in high esteem by their peers and other recipients left the unit shaking their heads and wondering what had happened to the awards system of the U.S. Army.

I had only one of those moments when they called the specialist that I had found to be more than a pain in the butt when I first came to the ETT. I was already offended because they chose to send him down to the hot spot on the Pakistan border instead of me and now they were awarding him a Joint Services Commendation Medal. I guessed it was because of the way he continually complained to any and all U.S. service members he encountered, along with our coalition partners; for this he was awarded a commendation medal for his efforts to keep everyone informed on the screwed up nature of our efforts in Afghanistan. If the Afghans had been giving out medals he would have surely qualified for a Taliban Sympathizers medal for all his bad mouthing of our host government's security forces.

I watched with interest to see if there would be any incidents when the CG showed up. There had been so much bad mouthing that I thought I might be able to get some memorable protest on tape for my video collection but the Oklahoma boys gritted their teeth and took their medals and the CG's remarks in a military manner. I'm sure the CO breathed a sigh of relief when, after a Canadian lunch, the CG got in his vehicle and left.

Then the word came down that all the infantrymen in the entire brigade had been awarded the CIB. That was interesting. There had been some pretty blatant attempts to fabricate shooting incidents for the purpose of qualifying for the CIB. When I heard about some of the incidents I just laughed to myself because these kinds of things always happen.

Now, however, the whole brigade got a blanket award and the ones who made false reports were mad. It was as if some of them didn't think that the others who did their jobs and never took the initiative to fake a firefight were as deserving as those who did. Then there was the residual suspicion that existed about some of the false reports being false reports and those involved not being eligible for the award like the others who had not been in combat either but had not filed any false reports.

The next issue to come up was the 'combat patch' or 'shoulder sleeve insignia-former wartime service' unit patch that is worn on the right shoulder of the army uniform.[118] The ETTs were part of the 45th Brigade which wore the 'Thunderbird' patch of the old 45th Infantry Division. The 45th Infantry Division had a proud history of wartime service in WWII and the Korean War and the general wanted everyone to wear the Thunderbird patch. The ETT guys were carrying a lot of hate towards the 45th Brigade, and the 'Cowboy' team, which had been OPCON to the SF in Gardez decided that they would wear a SF combat patch instead, which was an icon of 'cool' in this war. The CO was still trying to get through this deployment with a decent OER from the 45th CG so he asked them to change the patches for the awards ceremony that was coming up and for the big parade when they returned to their armory in Oklahoma. This caused a fair amount of grumbling as the personnel officers furiously continued to write the award citations that had to be completed by a certain date.

Our Replacements

The Indiana National Guard's 76th Infantry Brigade had been chosen to replace the 45th Brigade. The National Guard had sent several divisions to the Korean War but had mostly stayed home during Vietnam. While it had participated in the peacekeeping missions in the Balkans before this war, there had been three generations of Guardsmen who had never heard a shot fired except at the range once a year. Now the Guard was getting involved in combat operations. The regular army was still doing most of the heavy lifting in direct combat but the Guard was getting these kinds of support missions because the regulars were stretched thin. The ETT hadn't seen much combat by past war standards but they had seen some and as time went on they would see a lot more. This would gradually change the Guard from the inside as more and more of the soldiers deployed. The war in Iraq would see large numbers of Guard personnel and there would be more coming to Afghanistan.

One of them I was delighted to see was Max Krupp, a captain who had been in SOD and was one of the most squared away young officers I had had the pleasure of serving with. He was smart and a stud he was. When he commanded the LRSD he rented a civilian airplane with his credit card so that his unit could conduct a water jump that had been planned for a long time. The military aviation unit had canceled at the last minute so he acted in what could have been a career ender and they jumped right on schedule. We desperately needed officers like him and I was happy to see that he had grabbed the first combat assignment available which was to be a combat advisor to the ANA.

The CO got with his counterpart and began the process of prepping him for the hand-off by holding the fire hose to his mouth and turning on the water, just like what we had experienced in 2003. We had a bar-b-que to welcome the new guys and when a few of the Belgium paratroopers came sniffing around we forgave their rudeness by pointing to the DFAC. I videotaped this event, added it to my video production and I was finished. I had added the CO's music; it was by a band whose name matched the CO's politics and he was happy when I played the final product for him. I made a few copies and gave one to each team leader and told them that they could make as many copies as they liked.

Captain Max

Max left with one of the teams that were going up north. They took the Salang Tunnel, built by the Russians in the 1960s, the only route through the Hindu Kush Mountains that separate northern and southern Afghanistan. They were leading a convoy of ANA trucks and became separated from the trucks in the tunnel. When they reached the northern end they pulled onto the shoulder and waited for the ANA.

Suddenly an ANA truck came barreling out of the tunnel. According to witnesses, while shouting the alarm, Max kept his team leader from being crushed by pushing him to safety from where he was standing between two vehicles. The ANA truck slammed into the back of one of the WFRs parked on the side of the road. This WFR slammed into the one in front of it resulting in a chain reaction type catastrophe. Some off the trucks went over the side and took the long drop.

Max went down the slope and treated the injuries of the Oklahoma major they were replacing. He helped the injured major back up to the road and organized the treatment of the rest of the injured. After he and the ANA Kandak XO established security, he set up the SAT-COM and alerted a QRF from Bagram. He continued to treat the injured Americans and some Afghan civilians who had been parked nearby. Other Afghans were not so fortunate and several had been killed.

Max was back at Camp Julian the next day along with the healthy members of the new and old ETT. The old ETT guys were complaining that we hadn't done enough to help them. I thought that was a bit unfair, considering the distances and our assets. I asked some of them about the incident and they told me about Max's actions. I asked them to put him in for an award but they were so busy trying to get their own awards that it appeared that Max's actions got lost in the shuffle. This is typical in war but I wanted Max to be recognized because he was such a good officer and the military needs men like him to advance in rank for greater responsibilities.

He then volunteered for another team that was going south where there was fighting. We took a picture together and I bid him Godspeed. He went down to the northern part of Kandahar Province, the heart of the Taliban insurgency. It seems that they too were rising from the ashes of the conflagration that engulfed them in the winter of 2001. Max and his team of advisors lived with the ANA and conducted combat operations as part of Combined Task Force Bayonet. On June 17th, 2005 they were in contact for five hours with a large Taliban Force.

According to his citation and a letter from the commander of TF Gun Devil, a sub-unit of CTF Bayonet; he led an element of ANA soldiers against a numerically superior enemy force which resulted in the defeat and destruction of the Taliban element. He and his ANA soldiers were under close range enemy fire from AK-47s, 74s, PKMs, and RPGs. He maneuvered his Afghan soldiers using Pashto after his interpreter was hit. At one point he carried his interpreter on his back for 400 meters while under fire to the medics. He directed the fire of the helicopter gunships that eventually came to support his unit and he personally shot and killed or wounded several enemy fighters. Then he carried one of his dead ANA comrades back to the casualty collection point. All in all, Max had a helluva day. For this he was recommended to receive a Silver Star. Someone back in the States downgraded it to a Bronze Star with a V (Valor).

Ten years later I found out he had received this medal and an Army Commendation Medal for his actions at the Salang Tunnel. Since the ARCOM[119] is handed out for National Guard drill performance back in the States, it does make you wonder when you look at someone's fruit salad on their dress uniform. The only recognition that counts is your reputation with the ones you served with. Max's service as a combat advisor put him in an elite group of soldiers.

From the Korean War through Vietnam to Afghanistan and Iraq, few soldiers have experienced combat serving with foreign troops who don't speak English. Advisors don't have any of the amenities normally associated with the U.S. armed forces living with their counterparts in the field and as far as I'm concerned, they are the real heroes of the Afghan war.

Max came home and was offered the only SF position in the Texas National Guard. Already Ranger qualified, he went through Special Forces assessment and selection before attending and completing the SF Q (qualification) course. Texas was authorized to raise an SF company and he became an ODA commander, then the company commander. At this time he is the battalion commander of the only airborne battalion in the National Guard. I write this because men like Max are what are needed to rebuild the armed forces after the long war since 911. The best leaders are born and then developed by older leaders and by their own experiences. The service schools operate on the premise that anyone can be taught how to lead, producing leaders who can talk the talk. Max and those like him can also do that while they walk the walk.[120]

Combat Advisors

Combat advisors are the unsung heroes of the post WWII armed forces. From the Korean War until now, soldiers, in the main, marines, some sailors, and airmen have left the relative comforts of serving with their countrymen and lived, fought and died with native fighters. Forced by circumstances to perform heroic acts above and beyond the call of duty that are seldom known or recognized by their countrymen, they earn the respect and affection of these foreign troops they advise.

During Vietnam, the army didn't consider advisor service as important a ticket punch as service with American troops and I doubt if that's changed. Awards are usually minimal because of the lack of witnesses for an advisor's actions. Of course there are always exceptions. Recently, two combat advisors each received the Medal of Honor for their actions in the same battle.

A Few Less Things To Worry About

I took my video editing machine, monitor, and guitar to the post office at OMC-Alpha along with a rug I had bought at the bazaar. I brought the gifts that the interpreters from KMTC had given me and all my video tapes and DVD copies as well. The post office helped me get them packed and I sent all these things back to the states for $50, the modern war. I wanted to travel light but I still had several duffel bags full of equipment that had to be turned in at Fort Carson where we would de-mobe. I also had my rucksack and a large assault pack for my laptop and some books I wanted to read on the way home. The word was that we would travel by truck to Bagram from where we would fly home.

Patches

The ETTs were the last units of the 45th Brigade to return to the states. This of course infuriated them. The ETT guys felt that they had been screwed over by the 45th from the beginning. They were pouring over the regulations that could possibly suggest their having a wartime affiliation with the SF or even the 10th Mountain Division so that they wouldn't have to wear the hated 45th patch of their tormentors.[121]

The CO spent a lot of time suggesting, exhorting, and finally ordering them to take off the SF patches and sew the Thunderbird patch on their right shoulder before they got to Fort Carson where they would be under the gaze of the CG and other higher ups of the Oklahoma National Guard. Later, there would be a ceremony at their armory where a number of the WWII and Korean War veterans of the 45th Infantry Division would welcome them home. He didn't want to offend them by letting his men wear another unit patch.

CPT Max Krupp and author
(author photo)

"You'll get over it, Joe. Oncet I wuz gonna write a book exposin' the army after th' war myself."

(Bill Mauldin-45 ID-WW II)

Cruisin Kabul—One More Time

The call came to take all the gear that we didn't need for day to day activities and load it onto two trucks that would take it to Camp Phoenix and then to Bagram Air Base when it was time to go. We got everything loaded and then we found out that the escort vehicles had taken off without the trucks with our gear. Now the trucks were getting ready to leave without any escort. The CO was unwilling to trust sensitive items to the assurance of the Afghan drivers that they would take all these things to Camp Phoenix so another sergeant and I, the only ones available at the moment, each took a WFR, our weapons and radios and followed the trucks into Kabul.

The drivers had assured us through an interpreter at the gate that they knew how to get to Camp Phoenix so we each fell in behind a truck and drove through the city. It was dark and the traffic was heavy so before we were half way through town we were separated and on our own. Since I was the senior man I used the radio to tell the other sergeant to stay with his truck and I would stay with mine. I suggested that he keep track of any major terrain or manmade features enroute so he could describe where he was if he got lost.

If this was a hijacking we were in deep doo doo but since we couldn't communicate with the truck drivers I decided to trust that they were doing what they said and began to pray as we rolled through a part of Kabul that I didn't recognize. The drivers had taken a different route than we usually took because of the size of the trucks.

After what seemed like hours but was probably minutes I recognized a building that we passed every time we went to Camp Phoenix. I called the other sergeant on the vehicle radio and he said that he was just pulling into the camp. I breathed a sigh of relief and thanked the Lord for bringing us through one more moment of uncertainty. These were the kind of moments that I enjoyed the most in my army life but only after they were over. This one was over when I received and returned a salute from Rambo as I entered the gates of Camp Phoenix for the last time. We regrouped and made sure that the BDOC acknowledged our shipment and took responsibility for it. They signed a form which we gave to the S4 when we returned to Camp Julian later that night.

Bagram Bound

The next day it was time to go. I loaded my gear on one of the trucks and took a position facing out with my M4 at the ready. We drove through Kabul and I did my informal intelligence survey and waved to anyone who looked at me as we passed. Three out of four of the people I waved to waved back and about half of them returned my smile so I determined that we still had the support of the people in Kabul. How it was in the country side was difficult to say. The people of Mazar had been friendly but there was fighting down south so I was leaving my third war uncertain about its outcome.

We turned north and crossed the Shomali Plains. There had been a lot of fighting here over the years and there were large minefields still active along the road. Red rocks indicated mines in the area and white rocks declared the area was cleared. There were lots of both as we passed though on our way to Bagram. This had been the fruit and vegetable source for Kabul before the Jihad and it was famous for its grapes. Now it was being restored little by little and I could see vineyards some distance from the road and small villages that appeared to be flourishing. There were still many signs of the war, mainly derelict tanks and armored cars out in the open fields and the ruins of buildings on the side of the road. The traffic was moving quickly and there was a lot of it as more and more Afghans were able to get a car and a cell phone. Now they were doing what we do, driving and talking.

The village outside of Bagram Air Base was busy and expanding as more and more Afghans moved in to try for employment at the air base. As we passed through the security checkpoint several A-10 'Warthogs' flew over us as we were passing at least a hundred Afghan civilians lined up at the gate to be checked before entering the base.

The Blues

We drove down Disney Drive, named for a soldier that had been killed in an accident in February 2002 when the first troops occupied Bagram and made it the largest coalition base in the country. The trucks took us to a place near the old control tower where there were large tents with cots for use by soldiers like us who were waiting for a plane. There was a flight ops office nearby and a USO lounge with a memorial to Pat Tillman the Ranger who was killed by his buddies in yet another blue on blue incident in April 2004. [122]

Even with all the risk management procedures we were still having accidents and this was the third memorial to these that I had seen recently. I had stopped and saluted the memorial to the four Canadians at Camp Julian that had been killed by one of our F-16 pilots and now I was remembering these two U.S. soldiers who had stepped up when the war started only to be killed by accident. I know that the ones responsible for these deaths will never have another day that they don't think about something that only took a moment. The families of these fallen men will forever wonder why this happened because in war you can understand death inflicted by the enemy but not by your own.

Wars and Rumors

I moved into the tent with the rest and listened to the rumors and speculations regarding the date and time of our departure. One of the officers who knew about finance said that we would be out of here before the last day of August because they would have to pay us another month's hazardous duty pay and other benefits if we were here on the 1st of September. Whether that was the motive I couldn't say because the way we were spending money in Afghanistan it was hard to believe that anyone was concerned about fiscal responsibility. He was right though and on the 29th of August 2004 we boarded a C-17 bound for Peterson Air Force Base in Colorado Springs, Colorado.

While we waited for that day, the thin threads of civility finally began to unravel and I was asked to go talk to a major who had been awarded a bronze star at the awards ceremony and was now being berated by some of his team mates. The word 'coward' was being whispered around and this major looked pretty distraught. I knew he was a Christian so I read him some verses from the Bible and told him that everyone had different perceptions of people's actions in the war zone. I recommended that he confront the ones who had a problem with him and find out what they were complaining about. That, I told him, would settle the question of who was a coward and give them a chance to air their grievances. He seemed to be a little comforted when I was finished but I never knew if he took my advice. Things were going downhill with other members of the ETT but we would have to get to Fort Carson before they finally came to a head and exploded.

CIBs

The PX area had a shop selling uniform accouterments including cloth CIBs. I considered my situation and finally gave in and bought one with a star over the wreath because I had been awarded a CIB, second award, by the 45th along with every other member of Task Force Phoenix II who had ever held an infantry MOS. I hadn't read the regulation for issuing them but I was pretty sure that 99% of the Task Force didn't qualify but there were a lot of CIBs floating around in the Army that were suspect and I wasn't charged with conducting an investigation so I had one sewn onto my uniform for my work in the drug war and my service outside the wire with the 1st ETT.

I took my old CIBs and gave one to the CO. I could tell that he was not entirely comfortable with the way they had been issued to everyone who could spell infantry so I told him that he had earned his when he was in Gardez under fire. I had been around him for months and I was sure he would have held his own in a firefight. He was a fighter first with many other responsibilities and I was sure that he would have led the charge if need be. He told me he appreciated that and had it sewn on his uniform.

Customs

One of the last things we had to do was go through customs which was a huge pain in the butt. We had to drag all our duffel bags, rucksacks, and anything else out of our tent and over to another area where the U.S. Customs would inspect our things and mark the bags that had been cleared. We had to unpack our duffel bags and lay all our clothing and equipment on the ground while the customs guys waded through the piles and cleared our baggage. It was a joke and an idiot could have smuggled a machine gun but fortunately all our idiots had decided to forego any temptations and buy their machine guns when they got back to the states, thanks to warnings we had been given.

Leaving On A Jet Plane

We put our gear on pallets and waited for the manifest call. Some made one more run to the PX while the rest of us lay on our cots in the heat and tried to sleep as the fighter bombers roared down the runway nearby and some helicopters drifted across the Russian made tarmac. When the call came we staggered out to the airplane, a C-17 and climbed aboard. There was a passenger package in the cargo bay with comfortable seats. I grabbed one in the rear and settled in for one more flight from Mars to earth; my DEROS had arrived.

When the plane's wheels finally left the ground after a long run down the strip, the troops in the back let out a cheer which I suppose is obligatory for most soldiers returning from a war zone. I didn't cheer the first time I flew out of a war because I knew I deserved to die and I was carrying on my back the bodies of my men. I didn't cheer this time either because I didn't want to leave Afghanistan. As we winged our way back to the states I considered many things that happened during my time with the Afghan people.

The Americans and the Afghans

As much as I have mentioned many of what I perceived as failures of leadership and incompetence, my own as well as others on every level, I would be remiss however if I didn't say that the level of good being done by the American service personnel far outweighs the screw-ups that all veterans remember. Most American soldiers are basically good hearted. They're judgmental regarding cultural differences and most of them, like most people, are ignorant of history.

The National Guard personnel I deployed with were a quick study and grasped the main points of some complex tasks far outside the boundaries of their military training. Most of them made friends with the Afghans, especially their interpreters, and there are a number of Afghans now living in the states due to these friendships. American service personnel hope we will succeed in Afghanistan and the last thing I want to see is a repeat of April 30, 1975.

God and the Afghans

The Lord Jesus had healed my psychological wounds from my service in Vietnam and now I was considering the things I had seen in Afghanistan. The Afghan people had made a big impression on me. I found them to be normal human beings, very charming and likeable. They were just like us in that they were born with a father and a mother, they grew up in a family, they were given in marriage so the cycle could continue and eventually they would die like the rest of the human race.

The Afghans were caught in a religious trap. They were made in the image and likeness of God but they had been brought up under the oppression of Islam and they didn't know how to get out of it. The ones I met were against the Taliban but the Taliban were the best Muslims. The Bible warns us to stay away from those having an outward form of godliness, though denying its power.[123] The Muslims supposedly uphold Jesus as a great prophet but they deny His incarnation, death and resurrection; thereby denying that man must be regenerated with the divine life of God that is in Christ Jesus who is now the life giving Spirit. The culmination of our understanding of this mystery is expressed in the statement; God became a man to make man God in life and nature, function and expression, but not in the Godhead.

It is a testimony to our being wonderfully made that the Afghans are still able to function on a human level. I saw that life in Afghanistan has tied up the adult population with a psychological Gordian knot. And I knew from my own experience that the only way for them to be freed is by the sword, the Sword of the Spirit. I had seen how much the Lord needs His people to go to places like Afghanistan to bring them the good news of the Gospel. I had hoped that in some way I could be used for this purpose and now I was leaving.

Fort Carson—Again

We landed in Colorado Springs and were bussed to Fort Carson to begin our de-mobilization process. We each had a form which included a checklist of all the stations we needed to go through before we were issued a DD-214, Certificate of Release from active duty and for me, a plane ticket. The rest of the troops were moving as units back to their home stations but I was an orphan now; the services I had performed for the functioning of the headquarters were no longer needed. I still said hello to some of the ones I had worked with but I was becoming invisible as the time approached for us to go home.

One of the first things we did was go to the Central Issue Facility or CIF, and turn in the equipment they had issued us months before. Most of my equipment had remained in the duffel bags that I put it in before we left Fort Carson so it was as clean as it was when they issued it. The only item they wouldn't take was my IBAS. I had to take it back to the barracks and scrub the front of it. They took it the second time and gave me a copy of the huge printout that itemized every bit of the equipment I had been issued and returned.

After CIF I had to go to the medics and be checked out; it took the rest of the day, standing in one line after another, getting the blocks checked on my form. The Post Deployment Health Assessment form was as much concerned with our mental health as it was with our physical health. The word was out that if you complained of any mental problems you would be kept at Fort Carson for further evaluation so everyone put on a happy face. Since I had a referral to the Combat Stress Center at Bagram, they asked me several times if I was still troubled by whatever had caused me to be sent there in the first place.

I told them that my transfer had taken care of whatever problems I might have been experiencing and when they asked me what those were I told them to talk to my former CO because I had no idea, seeing how we weren't in combat. Since he had been through there a month before and was no longer available, they checked my block and I was good to go. The next station was where we got our DD-214s. I sat down and reviewed the one they had prepared for me, signed it and moved on to the last station which was travel; I picked up my plane ticket to Austin, TX for the morning of September 3rd.

Things Fall Apart

I went back to the barracks after lunch to lie down and fell asleep. I was awakened by the sound of someone screaming in the barracks. I jumped up and opened my door and stood there listening as someone down the hall continued to scream at the top of their lungs. I expected to see people running toward the source of the sounds but there were several people in the hallway, walking as if there was nothing going on.

I walked toward the room where the noise was coming from. It was one of the team rooms where the entire team was bunked together, to facilitate the final deluge of paperwork, the completion of citations for medals, necessary before they went home. The door opened for a minute and I saw a Captain who was the S1 for the team shouting out a litany of vituperation regarding various members of his team. One of the team members saw me looking in and came to the door.

Before I could finish asking if they needed some help he closed the door in my face. The shouting continued without pause so after listening for another minute, I walked outside to see if anyone else could hear. The man's voice was easily heard outside and was even echoing off the walls of the building across the courtyard. Soldiers were walking to and fro and no one seemed to be the least concerned about the sounds of distress coming from the team's room.

I went back to my room and listened carefully as I walked by. The gist of it seemed to be the captain's conscience objecting to the falsification of medal citations. He was the one who was supposed to be writing them and he sounded like he had had enough. I had a room to myself so I rented a car and ate on the town that night. When I returned to the barracks there was still a mild uproar but the shouting had stopped so I locked my door and slept till morning.

The Big Parade

The next day was a parade for the ETT where the 45th CG welcomed us home and told us what a good job we had done. By now I felt completely separated from the ETT so I stood on the sidelines and took video of the event. I did see my CO and told him about the captain's meltdown and he seemed like he was aware of the situation but didn't want to discuss it with me. I told him that he might want to get with the captain and help him through whatever was troubling him and that if he didn't these guys were going to have some pretty sorry reunions.

Some Good Advice

The last function we had to attend was a group therapy session by the chaplain. I wasn't much interested but it was mandatory so I joined the others in an auditorium. There had been some murders and suicides at Fort Bragg by some SF people when they returned home and the Army was trying to prevent the rest of its returning heroes from acting in a less than heroic way. I was pleasantly surprised with the chaplain's two common sense exhortations. He told us that our wives had probably done some things around the house while we were gone and to remember, "You weren't there!" and "Don't sweat the small stuff; remember, most of it is small stuff." That was the last time I saw everyone together.

Oklahoma-Where The Wind Comes Right Behind The Rain

That afternoon the 1st ETT was driven by bus to the airport for their flight to Oklahoma City and a homecoming celebration. I said farewell to the CO and to the others I could catch before they got on the bus and then they were gone. That evening after dinner I turned in my rent car and walked back to the barracks. They were quiet that night and I thanked the Lord out loud before I went to sleep.

Home Again

On the 3rd of September 2004, I caught my flight to Austin. I was in civilian clothes because I didn't want to attract any attention when I went through the airport. I knew that soldiers were getting better receptions this time around than we did coming back from Vietnam but I considered the war my own business and didn't want to share it with anyone. I called my wife and told her when I was coming in and she said she would pick me up at the airport. Richard, my son, was in school so she would come by herself.

I got off the plane and walked through the airport to the escalator that would take me down to the baggage pick-up and hopefully my wife. As I came down the escalator I was looking everywhere hoping to see her so we could grab my things and get out of there. I didn't see her anywhere and I was starting to imagine she had got the pick-up time wrong and I was going to have to play phone tag with her using the pay phone. I reached into my pocket to see if I had any quarters and in doing so looked down and there she was, standing at the bottom of the escalator where she had just stepped out from the wall where I couldn't see her. She was smiling at me and waving a small American flag. This was a nice touch on her part, homage to the end of one of my favorite movies, 'Heartbreak Ridge.' I wasn't any Gunny Highway but I appreciated the sentiment. She and the sisters she met with once a week had prayed me safely home through my last war with the U.S. Army and I was glad to see her.

Coming To The End Of The Road

I took 27 days of leave at home, the first time I had ever stayed at home on leave, and decompressed. I went through the video tapes of Afghanistan and I checked my email occasionally. What was I going to do? The Lord had shown me his love for these people, two had gotten saved, and here I was back in the states with nothing to do. On top of everything else I was facing my retirement from the only thing I had ever wanted to do my whole life. I went into denial.

The Afghans
(1 ETT photo)

The Afghans
(SPC Marriner photo)

The Afghan National Army
(Major Cowan photo)

For the weapons of our warfare are not fleshly but powerful before God for the overthrowing of strongholds,
2 Corinthians 10:4

Chapter Eleven: October 2004-March 2005

The Unintended Consequences Strike Again

I went back to work in October and reported to Major M. He was the TAG operations officer in Afghanistan and was now the CD Operations Officer. He told me that some things had changed while we were in Afghanistan. The reorganization had somewhat stalled because of a number of IG complaints against the CD program. The crux of the matter was the variation in rank caps for the various activities in the CD program.

Putting a cap on promotions was seen as the way to keep the number of personnel we put in the field to support law enforcement from shrinking. The more we had to pay each person the fewer people we could afford to employ. The Army Guard had improved its promotion system, taking it out of the 'good old boy-brother-in-law' method that it had used since voting for the leaders was done away with. Now there was a measurable criterion used to determine an order of merit promotion list for each rank in each MOS in the Army Guard; who knows how they do it in the Air Guard?

The people who were in the CD program got more promotion points than most of the traditional guardsmen because they were free to attend military schools and work on correspondence courses along with other activities that produced points. This put them at the top of the promotion lists each year. If the CD program didn't cap the ranks we would have 10 colonels and 20 sergeants major working for the same money that we had been getting for 300 people.

In the early days of the program the intel guys, claiming that school trained intel analysts were rare, managed to get the leadership to cap the enlisted rank at E-8. The cargo inspectors didn't have an MOS[124] associated with their duties so they were capped at E-6. This had been the accepted practice for ten years. The whole reorganization process, on which I had labored long and hard before I went to Afghanistan, was started because some of the cargo inspectors were trying to move into intel slots so they could get promoted.

Now the law of unintended consequences finally kicked in. Tony, my career enabler and mentor made a complaint to the IG after CD prevented him from taking a promotion to E-9 for the third time. Several of the cargo inspectors had also complained. The IG looked at the whole picture and determined that because most of the intel analysts were white and most of the cargo inspectors were Hispanic, the program had been racist in its personnel practices.

Since I was deeply involved in the reorganization I reviewed in my own mind all the discussions about rank structure and caps and I honestly couldn't recall any thought of favoring one race over another. The intel program had grown faster than any other CD program and eventually we ran out of school trained analysts and a number of our people were working with the cops as analysts who had no school training. The problem with rank cap was one of the things we had tried to alleviate with the reorganization and since the cargo inspection mission was being done away with I hadn't anticipated anything like this.

LTC Randy Davis had been blamed for favoritism regarding Tony's promotion and had been transferred out of the program and was now at the academy. Tony got promoted to sergeant major. There were several other promotions and personnel changes as a result of these investigations. COL Russ was looking kind of shell-shocked when he welcomed me back and he couldn't even tell me why Colonel Davis was gone. When I went back to work I asked Air Marshal Jim what was happening with the reorganization. He told me that it had gone well for a while but had stalled after Brian had left for Iraq. I started to get into it but I could tell that something was going on.

Time To Go

A few days after I came back to work an issue arose about the personnel evaluation system that I had developed before I left. It seems that the people in the field didn't like the criteria with which they were to be evaluated. I went to Major Shawn and fully expecting him to agree with me. I told him that the headquarters needed to hold the line on this, in fact, why was it even a matter for discussion? He hedged around a bit and then he told me that the old regime, including me was under suspicion of favoritism, and that he was there to clean house. I became very angry for a short period of time and then I came to my senses and told him that I was retiring and that it would be better for me to focus on retirement and let the new guys learn the new system.

A Fool's Errand

Then I was told that I had to go to Washington D.C. to the Guard Bureau for a week long class on CMIS. That was too much. I had just been told that I was no longer of any use because I was part of the old regime that had conspired to practice racist personnel policies. Now they wanted me to go to DC and get a class on the hated CMIS program that had evolved into a monster. I went through the entire chain of command explaining why I was the last person they should be sending to this class. Beside the fact that I despised the CMIS program and the people who had inflicted it on us, I was supposed to retire in a couple of months so why would they not send someone else who would be staying for a while?

My objections to this idea fell on the deaf ears of my boss and Col M who ran the program. Finally, a sergeant who knew what was really going on told me that the class had been paid for with end of the year money[125] and they didn't have anyone else in the program that was free to go so they were sending me to fill a seat. I was angry but I knew how to follow orders. If they wanted me to go to Guard Bureau two months before I retired I'd go; when I got there I'd give the Bureau some feedback from the field.

Feedback From The Field

I flew to D.C. with several others from the HQ. We were billeted in Crystal City at a hotel which was connected to the Guard Bureau. I showed up for the class that was given by a female air guard sergeant who was quite pleased with herself and her superior knowledge of the CMIS program. I tried to be civil but within the first hour she had identified me as an adversary. I tried to contain my disdain for the CMIS program but after she gave me a few flippant answers to my questions regarding the necessity for certain data that no one could possibly be interested in, I declared war.

The people who ran the CD Management of Information System (CMIS) had continued to add more and more pages and fields for data input. We had to train everyone in CD to do their own input because the HQ data entry person was totally overwhelmed by the additional requirements that had been added each year. Every time someone in congress mentioned anything about the CD statistics they added more fields and instead of CMIS allowing Guard Bureau to pull the stats they needed, we were getting more and more demands from them to produce huge reports with the most obtuse details about things that no one but an idiot would need to know.

I now took every opportunity she presented to denounce the insatiable demands on our time by the CMIS staff. They had required us to input enormous amounts of data while assuring us that they would be able to create summaries for the Bureau and the Congress. Instead, they demanded that we produce the reports that they had assured us that CMIS would produce. After several exchanges she questioned my integrity so I let her have it. I gave her a detailed summary of the report demands from NGB during the past couple of years. In the class there were a number of other CMIS workers from states like Texas and they verified what I was saying. By the middle of the second day of this ordeal she was having to explain the unexplainable to them as well and she was very angry with me.

At the break she told me that a LTC in the CD directorate wanted to speak to me. I knew that she must have complained to her higher ups and now they were going to squash the old sergeant from the hinterlands that dared to question what they were doing at the center of government. I followed her directions and walked through the warren of cubicles until I found the LTC's office which occupied a coveted position on the wall and it had a window.

B-L-U-F

The LTC asked me in and offered me a seat. As I sat down he asked me what my problem with the CMIS class was. I liked that, bottom line up front, so I told him. I gave him a brief summary of my history with Counterdrug and my observations of the evolution of the CMIS program. I told him that it had been using valuable man hours at our end to gather and report the information that the Guard Bureau kept insisting the program would do for us but never did.

He wasn't expecting to deal with someone who had been on the program longer than he had so he tried to placate me by telling me how he had been out in the field before he came to the headquarters. I'm sure he was doing hard work out in the field but I doubted that he was doing any data entry or surveillance so I gently informed him that it didn't seem that we had much in common other than that we both drew a paycheck from the U.S. government. He seemed disappointed that he couldn't lead me to some common ground where he could gradually gain my submission to the superior logic that existed in the rarefied air of the CD Directorate of the National Guard Bureau. Now he took a different tact and asked me what was wrong with CMIS.

Bottom line up front again, so I gave it to him. Fifteen minutes later he had had enough. He took notes as I went down the list that had coalesced in my mind during my discussions with the Air Guard sergeant who taught the class. The LTC thanked me for my feedback from the field and hoped that I could understand the intense pressure they were under in the Guard Bureau. I assured him that I did and that I hoped that with the information that I had given him they could use it to resist the urge to pressure the people in the field with work that the Bureau should be doing.

He thanked me again but didn't extend his hand like we do in Texas at the end of a conversation. I figured it was a D.C. thing or maybe an officer/enlisted cultural barrier, but either way I didn't care. I knew that our conversation wasn't going to have the slightest effect on the CMIS beast and I wondered if it would get back to Texas. I shouldn't have though because a detailed report about the master sergeant who dared to question the efficacy of the CMIS program, and by insinuation the CD Directorate of the National Guard Bureau, was on Col M's desk before I left D.C.

The Mall and The Wall

I spent the last afternoon we were there walking around the capitol. I saw the new WWII memorial on the mall and I had one more nostalgic visit to the Wall where I read the names of my comrades who gave their all on June 17th 1967 during Operation Billings. After the Wall I took the subway back to the hotel, scored a pizza to go and ate it while watching planes making their approach to the airport where I had worked in '68. The next day I caught a flight back to Texas.

What Did You Say?

When I came in to work on Tuesday I was told to report to Col M. One of the officers in the HQ who knew me asked what I had done in D.C. to cause such an uproar. Forewarned is forearmed so I knew that Col M wasn't calling me in to reminisce about our long history with CD. He is a decent man and slow to anger; I had never seen him demonstrate anything but sorrowful disappointment when he discovered that members of his organization had behaved in a less than professional manner.

I entered his office and reported to him. He invited me to sit down and proceeded to tap dance all around my visit to NGB and what had I said to them that resulted in the nasty letter that he had received from the chief of the CD Directorate. I gave him a few details of my conversation with the LTC and when I was done he thanked me for coming in and asked me to talk to the Sergeant Major of the CD Task Force. I went to his office as soon as I left the colonel's office and he gently chided me for my conversation with the CMIS people at NGB. The last thing he said was that I shouldn't do it again. I assured him I wouldn't, seeing how I was retiring in a few months and left his office.

The Lord's Move

In December my wife and Richard and I went to a Bible training in Anaheim. We took a Chinese brother named Raymond with us. He was a communist from mainland China who had come to the U.S. to get his PhD in nuclear physics. He had met some Christian students on the UT campus, heard the gospel and received the Lord Jesus into his spirit. He told me about life in communist China and how he came here. I praise the Lord that he was able to move men all over the earth for his purpose and that the boundaries of the nations and the times of their habitation are set by him.[126] I hoped that some of the Afghans would be able to come to the U.S. someday.

The End of That Road

When I returned from my leave I was finally forced to consider my retirement. I had never given much thought to the subject. I didn't join the army for benefits like retirement but now I couldn't avoid it any longer. The CD program required us to retire when we had 20 years of active duty. The army had a mandatory retirement age of 60 years old. I was 59 and I considered resigning from the CD program and joining a M-Day unit but that wouldn't accomplish anything. Another old master sergeant was not what most units were looking for. The personnel people had discovered six months of active duty time when I went to basic training back in 1963 and told me that I could have retired when I was in Afghanistan. They set up appointments for my retirement physical and the retirement transition program.

I made yet another trip to BAMC, in the new building now at Fort Sam Houston in San Antonio. I had been to Fort Sam many times during the past fifteen years. I was gone over by the medical people and at the end a Captain who was a PA went over the results of my physical and the long list of physical problems I had encountered during my time in the military. He was wearing an SF tab so I asked him if he was an SF medic and he said that he had been an SF ODA team leader who went to PA school because he wanted a change. He asked me why I had been in and out of the military since 1963, more than forty years. I told him it had gotten into my blood and I couldn't get rid of it.

When he got to PTSD I had written that I had had it but now it was in remission. "What do you mean by that," he asked. I told him about how the Lord Jesus had healed my soul from the effects of being an infantry sergeant in combat in Vietnam. "Every time I had a flashback, I would call on the name of the Lord Jesus," I told him, "Oh Lord Jesus," I called with a soft voice. "In three years I was healed so my PTSD is in remission. I can still remember much of what happened in Vietnam but I'm no longer tormented by it. If I stop calling though, it will come back eventually." He looked at me like I was crazy and hurried to complete the interview. I marveled at God's wisdom in Christ and I realized that even though these things are written in the Bible, even those who read them have a hard time entering in. As I left BAMC I called on the Lord and thanked him.

My next appointment at Fort Sam was a retirement transition seminar that I was required to attend. I took my wife as they recommended and sat there in a daze while she took notes. I just couldn't get into all the programs and websites for retired soldiers that they were recommending. I did pay attention when they told me that many retired people died a few years after they retired if they didn't have anything to do. I wanted to draw the money they said I would be getting every month for as long as I could; I had worked much longer than the eight hour days that many of the REMFs enjoyed so I considered my retirement pay as a kind of overtime.

At the end of the day we took all the brochures and my wife's notes and as we drove back to Austin she asked me what I had learned. I told her that I needed something to do and that I was going to write a book exposing the army just like the dog faces in my favorite Bill Mauldin cartoon. I asked her what she got out of it but she drew a blank so we never did any of the things that they suggested.

Taking Care Of The Troops

I had to go back to Fort Sam one more time and that was to get a final DD-214[127]. While I was waiting for the man to finish adding all the information needed to the form, I mentioned that it appeared that if I had fourteen more days I would retire at the next pay level which would mean about $200 more a month. He stopped everything and made a quick check. He discovered that I was right.

I told him that I had mentioned it to the personnel sergeant at CD and he told me that there was nothing he could do. The man picked up his phone and called the personnel officer at Camp Mabry and told him that they should let me stay in for fourteen more days. When the officer at the other end hesitated, the man told him that he would call the Adjutant General to get this straightened out. That helped the officer decide to go with the man's recommendation and I was told to come back in two weeks when we would complete my 214. That was the final and one of the best examples of 'taking care of the soldier' I had ever seen and he was a civilian.

When I returned to CD that day, the Colonel put me on admin leave so I could stay home and not have to sit around the HQ doing nothing every day. I cleaned out my desk and waited until most of the people had left. Then I took my papers, books, and data disks out to the car and drove home. No parades, no farewells— those things were not being done at Camp Mabry for sergeants. I drove through the gate that night just like I had for the past fifteen years. I drove south on Mopac still not in the realization that this was the last time.

I did however invite some of the men I served with to come to my house for hamburgers during their lunch hour to thank them for helping me do my job. I told them that I had been waving the ethical flag during my time at CD because we had to continually remind the troops of our responsibilities regarding our code of ethics. "But that's not the same as believing in what the Lord Jesus did, dying on the cross for our sins and resurrecting as the life giving spirit to regenerate us unto a living hope." The CD sergeant major attended although I hadn't invited him and he surprised me by affirming everything I had just said. At the end of the two weeks I returned one more time to Fort Sam where I signed several copies of my final DD-214 and that was it. I was done.

What Next?

I told my wife that I wished I could have stayed longer in Afghanistan and she told me that she would go with me where we could sleep beneath the stars under a blanket of snow. Unfortunately by that time the Taliban had been reconstituted and the insurgency was growing in intensity; Afghanistan wasn't issuing any tourist visas. I had kept my army email account and one day I got an email from a sergeant in Afghanistan. He said that an Afghan interpreter named Matt had asked him to contact me. This was the start of me and my wife becoming much closer to becoming Afghans.

The Afghans
(SPC Marriner Photo)

The Afghans
(SPC Marriner photo)

The Afghans
(SPC Marriner photo)

Epilogue

Three years after I returned home, one of the interpreters, Matt, came to live with my family. After several years we were able to get his family here and we have lived next door to them since 2010. Their five children have become my grandchildren.

My son Richard and his wife McKenna and their two children Hudson and Indigo also live with us. Richard became a Paramedic, joined the Guard, went airborne and will soon be a SOCOM Medic

The soldiers I served with went on to more deployments in Iraq, Afghanistan and other areas of the world where we fight. There hasn't been a lot of interest in reunions and though a few of us try to keep in touch on Facebook, it will be another decade when the younger troops will try and reconnect with their wartime comrades.

The issue is still in doubt in Afghanistan. Most of my Afghan friends are glad that the U.S. hasn't abandoned them and they hope, as I do, that peace will come to their country.

Appendix 1: Statistical Results

Training Assistance Group II (December 2003-August 2004)

1. The TAG contributed to the training and graduation of over 9600 soldiers at KMTC.
2. Increased KMTC's training capacity from 2 to 4 battalions (Kandacs) per rotation, by expanding infrastructure, billeting, DFAC operations and training ranges.
3. Increased operational firing ranges from 13 to 44 enabling the ANA to train more personnel.
4. Increased range control operations from 8 hours per day to 24 hours per day, while transitioning from a U.S. ran operation to an ANA ran operation, enabling the ANA to achieve compliance with critical safety requirements and communications from only 40% to 95% coverage.
5. Implemented the use of organic ANA transportation assets to replace contract transportation saving $90,000 per month.
6. Assisted OMC-A with the implementation of a cook training program for ANA soldiers to replace contractors, saving over $3.0 million per year, while eliminating major sanitation problems in the dining facility.
7. Implemented an installation property accountability system improving the ANA's ability to control all equipment issued to them for use at KMTC.
8. Provided outstanding leadership, oversight and mentorship to critical ANA functions at KMTC to enable them to be self-sufficient. [128]

Training Assistance Group II-III (November 2003-August 2005)

From November 2003 through August 2005 the 136th Regiment (RTI) with soldiers from Texas, Rhode Island, Alabama, and California deployed to Afghanistan as part of Coalition Joint Task Force Phoenix. Together they formed and staffed the Afghan National Army (ANA) Training Assistance Group (TAG) at the Kabul Military Training Center (KMTC).

The unit utilized its military instructor skills and staff experience to coach, mentor, and train over 1000 Afghan officers and NCOs responsible for running basic training, advanced NCO training, and officer training for the Afghan Army. The training staff consisted of soldiers from the United States and nine other coalition armies to mentor and empower the Officer and NCO Corps to become an integral part of the day to day training as well as the decision making process.

The training staff assisted the Afghan Army to graduate and incorporate over 20,000 enlisted soldiers and 1000 officers into the new Afghan Army. Additionally, they helped train over 3000 soldiers in field artillery, ordnance, quartermaster, armor, reconnaissance, mechanized infantry, engineer, signal, and medical advance training skills.

The garrison staff coordinated with the Army Corps of Engineers and local contractors to improve and expand the infrastructure at KMTC. By building new and renovating existing structures, KMTC was able to expand its training base from 2 training battalions to 6, training over 3200 soldiers at the same time.

The garrison staff also expanded the number of operational firing ranges from 14 to 44, expanded the force protection posture of the training facility, increased the training cadre staff by 30%, and incorporated a supply accounting system that maintained accountability of all equipment and resources donated to the Afghan Army.

All in all the Training Assistance Group effectively mentored and trained the ANA leadership, enabling them to graduate over 20,000 soldiers to help protect the security of the nation, incorporate democratic elections, and to increase their economic prosperity. [129]

Appendix 2: 10 Week Basic Training Summary

Zero Week – Zero Week is soldier in-processing which consists of basic soldier information, medical exam, KMTC orientation, basic clothing issue (CIF) and an emphasis on drill and ceremony. Physical fitness is introduced and stays current throughout the course.

Weeks 1-2 – are conducted in a round robin type format. The primary focuses are Basic rifle marksmanship, CTT training, individual movement techniques; camouflage self and equipment, Law of War, noise and light discipline and more drill and ceremony. Physical Fitness Test is conducted each week for each company.

Week 3-4 – is the first week in which the entire Kandak trains together. The focus is on buddy team movement techniques, with a Blank Fire Exercise culminated by a life fire exercise. More drill and ceremony.

Week 5 – is conducted as a round robin format for the first couple of days. Soldiers are taught Fire Team Movement techniques and PKM & RPG familiarization. The week is culminated by a live fire exercise with the entire Kandak and a 12K road march. More drill and ceremony.

Week 6 – Soldiers are taught Squad duties and responsibilities and Squad Team Movement techniques. The week is culminated by a live fire exercise with the entire Kandak and a 15K road march. More drill and ceremony.

Week 7 – Soldiers are taught basic anatomy, medical tasks and communications. The main focus is on react and break contact. There is a blank fire exercise conducted culminated by a live fire exercise. More drill and ceremony.

Week 8 – Week 8 is called professional week. Soldiers are orientated to the different functions of the army. They are shown communications, quartermaster, transportation, artillery, mortars, engineering, and medical. More drill and ceremony.

Week 9 – Week 9 is Platoon Week. It is an extended week which actually runs into week 10 by 3-4 days. The soldiers are instructed in Platoon duties and responsibilities, movement techniques, movement to contact, ambush techniques and platoon raid. More drill and ceremony.

Week 10 – This is graduation week. This week is when the Officers and NCO's are reintroduced back into the Kandak. Soldiers practice drill and ceremony for the entire week. They are given there final issue of gear and equipment. (If Available) They participate in a graduation ceremony are turned over to their follow on commanders and are transported to their new brigade. End of drill and ceremony.[130]

Glossary

Bazaar - Afghan market place
Burka - Pakistani word for Afghan women's clothing known as a Chadori
Buzkashi - Afghan Polo
Camelback - Water bag carried like small backpack with a drinking tube
Chadori - Afghan women's clothing - Sometimes called a Burka in Pakistan
Dari - Afghan version of Farsi, the language of the Persians-Iran
Durdee - Bread or a meal - Pashtun word for Nan, the bread of Central Asia and the Middle East
Eid al Adha - Muslim Feast of Sacrifice - The 'Big' Eid
Fobbits - Term used to describe the REMF's of the War on Terror - Inhabitants of the FOB
Gullbuddin Hekmatyar - Mujahideen Commander during war with Russia, now leaders of the HIG
Hazaras - Afghans believed to be descended from Genghis Khan's army.
Hazarajat - Home of the Hazaras - Bamiyan Province - Central Afghanistan
Helipad - Permanent Helicopter Landing Ground
Kabul Krud - Diarrhea
Kandak - Afghan Battalion
Karakul - 'Karzai' hat
KHAD – DRA Intelligence Service
Kuchi - Afghan nomads
Pashto - Language of the Pashtun people-Large ethnic group in Afghanistan and Pakistan
Pakol -'Massoud' hat
Picatinney Rail - M4 accessary - used for mounting various sighting devices
Pogues - Persons Other Than Grunts - Derogatory Term used by the infantry to describe other service members
Rabbani, Burhanuddin - Former President of Afghanistan - assassinated by Taliban suicide bomber in 2011
REMF Rear Echelon Military Forces
Spectre - AC-130 Gunship - carries a 105mm artillery cannon, 40mm automatic cannons, and various sensors
Sharia - Islamic law

Bibliography

The following books were found by me to have useful information for someone wishing to know more about the people, places, and history of Afghanistan and Pakistan. As an intelligence analyst I have tried to read as many books and articles about whatever subject I'm studying. Eventually it is possible to come up with some theories about what might have happened and why. When you do this to study Pakistan and Afghanistan be prepared to have your theory's demolished by the inhabitants of those troubled places. Unless it's a conspiracy theory, you're wrong.

Books

The Holy Bible-Recovery Version Old and New Testament Anaheim, CA: Living Stream Ministry 2003-All Bible verse quotes come from this excellent study Bible.

Adamic, Ludwig W. ***Historical Dictionary of Afghanistan-2nd Edition*** Lanham, MD: The Scarecrow Press 2002-A good source of Afghanistan history

Amir Habibullah ***My Life: From Brigand to King*** London: The Octagon Press 1990-An interesting autobiography of Bacha Saquo as told to Jamal, his companion. The only Tajik to be King of Afghanistan.

Blaber, Pete ***The Mission, the Men, and Me-Lessons From a Former Delta Force Commander*** New York: Berkley 2008-Good lessons on how to think-history of what really happened on Operation Anaconda and why.

Caroe, Olaf ***The Pathans*** New York: St. Martin's Press 1958

Coll, Steve ***Ghost Wars*** New York: The Penguin Press 2004-Well researched, a good source of history.

Christ, James ***The Bone Yard*** Amazon DS: Battlefield Publishing 2011-The battle of the Bone Yard

Crile, George ***Charlie Wilson's War*** New York: Atlantic Monthly Press 2003-Congressman at war

DePuy, William E. ***Changing an Army-A Oral History…*** Washington, DC: U.S. Army Center for Military History 1988-An oral history of the methods used to rebuild the Army after the Vietnam War

Dupree, Louis ***Afghanistan*** New Jersey: Princeton University Press 1973

Dunnigan, James F. and Nofi, Albert A. ***Dirty Little Secrets of the Vietnam War*** New York: St Martin's Press 1999-Some things we still don't like to talk about.

Elliot, Jason ***An Unexpected Light-Travels in Afghanistan*** New York: Picador 1999-Adventures of a young British traveler on his two trips to Afghanistan in 1979 and at the end of the eighties

Farivar, Masood ***Confessions of a Mullah Warrior (Uncorrected Proof)*** New York: Atlantic Monthly Press 2009-Memoir of a mujahideen fighter against the Russians

Feifer, Gregory ***The Great Gamble -The Soviet War in Afghanistan*** New York: HarperCollins 2009 A history with personal accounts of Russian troops

Friscolanti, Michael ***Friendly Fire-The Untold Story of the U.S. Bombing that Killed Four Canadian Soldiers in Afghanistan*** John Wiley and Sons, Canada 2005-Pilots feel the need for speed, take some, and bomb our allies

Fury, Dalton [pseud] ***Kill Bin Laden*** New York: St Martin's Press 2008-Delta Force officer explains why they weren't able to 'Kill Bin Laden' at Tora Bora.
Ghobar, Mir Gholam Mohammad ***Afghanistan in the Course of History - Volume Two*** Alexandria, Virginia: All Prints 2001-Afghan scholar's controversial history of Afghanistan
Gole, Henry G. ***General William E. DePuy-Preparing the Army for Modern War*** Kentucky: University Press of Kentucky 2008-Biography of Gen William DePuy-Rebuilding the U.S. Army after Vietnam
Grau, Lester W. [editor] The Bear Went Over the Mountain - Soviet Combat Tactics in Afghanistan Ft Leavenworth, KS: FMSO 2005-Russian officers experiences in Afghanistan 1979-1989
Grau, Lester W. And Gress, Michael A. [editors] ***The Soviet - Afghan War How a Superpower Fought and Lost*** Lawrence, KS: University Press of Kansas 2002 The Russian post-mortem on Afghanistan
Griffin, Michael ***Reaping the Whirlwind - Afghanistan, Al Qa'ida, and the Holy War*** Stirling, Virginia: Pluto Press 2003-Good history.
Hopkirk, Peter ***The Great Game - The Struggle for Empire in Central Asia*** New York: Kodansha America 1992-Excellent history - Her Majesty's Secret Service in the Hindu Kush.
Hosseini, Khaled ***The Kite Runner*** New York: Riverhead Books 2003-Best selling historical novel and movie of the same name. A help to understanding Afghan culture and a great read.
Isby, David ***Afghanistan-Graveyard of Empires-A New History of the Borderland*** *New York:* Pegasus Books 2010-Focus on Pashtunistan and all that it entails.
Jalali, Ali Ahmad and Grau, Lester W. ***The Other Side of the Mountain-Mujahideen Tactics in the Soviet-Afghan War*** Quantico, VA U.S.M.C. 1995 Oral history of mujahideen-cited by U.S. SOF as a key to Taliban tactics.
Kipling, Rudyard ***Kim*** Mineola, NY: Dover Thrift Edition 2005-Classic novel of the Great Game.
Macintyre, Ben ***The Man Who Would Be King-The First American In Afghanistan*** New York: Farrar, Straus, Giroux 2004-Could be the real model for Kipling's classic by the same name.
Magnus, Ralph H. and Naby, Eden ***Afghanistan - Mullah, Marx, and Mujahid*** Boulder, Colorado: Westview Press 2002-Excellent history of the Afghan communist movement.
Masters, John ***A Bugle and a Tiger*** New York: Viking Press 1956-Excellent memoir of a young officer with the Gurkhas on the Northwest Frontier of India prior to WWII.
Meyer, Karl E. And Brysac, Shareen Blair ***Tournament Of Shadows-The Great Game And The Race For Empire In Central Asia*** Washington D.C: Counterpoint 1999-Another view of the Great Game
Michener, James ***Caravans*** New York: Random House 1963-Adventure novel of post WWII Afghanistan, a good read with a section regarding the U.S. efforts in Helmand Province.
Miller, Charles ***Khyber - British India's Northwest Frontier-The Story of an Imperial Migraine*** New York: Macmillan 1977-Excellent history, Mr. Miller has a sense of humor.

Moore, Robin ***The Hunt For Bin Laden-Task Force Dagger*** New York: Random House 2003- Sensational book describing the SF (Science Fiction) operations early on in the invasion of Afghanistan.
Schroen, Gary C. ***First In - An Insider's Account of How the CIA Spearheaded the War on Terror in Afghanistan*** New York: Ballantine Books 2006-A 60 year old man in the retirement process is sent to war.
Stewart, Rory ***The Places in Between*** New York: Harcourt 2004-He walked across Afghanistan in 2002
Sykes, Sir Percy ***A History of Afghanistan Vol II*** London: Macmillan 1940-A British version of Afghanistan history
Tolkin, J. R. R. ***The Hobbit*** New York: Ballantine 1966-I loved this book in 1968. Couldn't finish the trilogy.
von Tunzelmann, Alex ***Indian Summer-The Secret History of the End of an Empire*** New York: Picador 2007-A great history of the Partition
Wright, Donald P. et al ***A Different Kind Of War-The U.S. Army in Operation Enduring Freedom Oct 2001-Sep 2005***-Fort Leavenworth, KS: 2010 Combat Studies Institute Press Official History

Manuals and Regulations

TC 25-20 DA SEP 93 ***Leaders Guide to AAR's***-How to do an AAR
FM 100-14 Risk Management APR 1998-How to cross the street safely, don't forget your reflective belt.
FM 22-100 JUN 1973 ***Military Leadership***-Published at the end of the U.S. combat mission in Vietnam-Uses Abraham Mazlow's Hierarchy of Needs to illustrate human nature.
FM 22-100 OCT 1983 ***Military Leadership***-Uses many historical events to illustrate leadership principals
FM 101-5 MAY 1997 ***Staff Organization and Operations***-Home of the dreaded MDMP and other Jedi mind warping exercises
AR 600-8-22 ***Military Awards*** **FEB 1995 - DEC 2006**-Who gets the cool looking badges
AR 670-1 SEP 2003-FEB 2005 ***WEAR AND APPEARANCE OF ARMY UNIFORMS AND INSIGNIA***
Where does all this stuff go on my uniform
ALARACT 055/2007 AR 670-1, 3 FEB 2005-DA Message dealing with combat patch controversies

Papers and Articles

Grau, Lester W. ***The Takedown of Kabul-An Effective Coup de Main*** Leavenworth Combat Studies Institute 2002
Lewis, LTC Troy D. ***Training the Afghan Army-A Monograph*** School of Advanced Military Studies-USCGSC, Fort Leavenworth, KS 2011-A SAMS Jedi Knight paper cgsc.contentdm.oclc.org/utils/getfile/collection/p4013coll3/id/2706

Makhzan-i-Afghani (The History of the Afghans-1600's
http://kblibrary.bih.nic.in/Vol06/Bp193.htm
The source of the claims that the Pashtun people are descendants of the children of Israel
Morrison John E. and Meliza, Larry L. ***Foundations of the After Action Review Process*** Special Report 42-U.S. Army Research Institute for the Behavioral and Social Sciences-History of the AAR
Stewart, Richard ***War in the Persian Gulf*** Washington D.C.: U.S. Army Center for Military History 2010 www.history.army.mil/html/books/070/70-117-1/index.html Official history.

End Notes

1 Henry G. Gole General William E. DePuy-Preparing the Army for Modern War (University Press of Kentucky 2008)
2 John E. Morrison and Larry L. Meliza Foundations of the After Action Review Process (Special Report 42-U.S. Army Research Institute for the Behavioral and Social Sciences)
3 Leaders Guide to AAR's-TC 25-20 DA SEP 93
4 Romie L. Brownlee and William J. Mullen III Changing an Army-An Oral History of Gen William E. DePuy-Page 187 (Center for Military History Publication 70-23 1988)
5 Page 135 (Military Leadership-FM 22-100-OCT 1983)
6 The land of the Afghans
7 Ch. 1, Art 4, Afghanistan Constitution
8 Ralph H. Magnus and Eden Naby (Afghanistan-Mullah, Marx, and Mujahid) Page 10 Westview Press 2002
9 Ch. 1, Art. 16, Afghanistan Constitution
10 Ezra Chapter 1 (Bible-Old Testament)
11 Makhzan-i-Afghani (The History of the Afghans-1600's) http://kblibrary.bih.nic.in/Vol06/Bp193.htm
12 Charles Miller (Khyber-British India's North West Frontier-The Story of an Imperial Migraine) Pages 3-16 Macmillan Publishing Company 1977
13 Ludwig W. Adamec (Historical Dictionary of Afghanistan) Page 87 Scarecrow Press 2002
14 Ralph H. Magnus and Eden Naby (Afghanistan-Mullah, Marx, and Mujahid) Pages 24-32 Westview Press 2002
15 Karl E. Mayer and Shareen Blair Brysac (Tournament of Shadows-The Great Game and the Race for Empire in Central Asia) Page 113 Counterpoint 1999
16 Hopkirk, Peter (The Great Game) Pages 22-23 Kodansha 1992
17 ibid Page 25
18 ibid Page 123
19 Ben Macintyre (The Man Who Would Be King-The First American In Afghanistan) Farrar, Straus, and Giroux-Macintyre Books 2004
20 Sir Percy Sykes (A History of Afghanistan Vol II) Pages 173-177 Macmillan 1940
21 Alex von Tunzelmann (Indian Summer-The Secret History of the End of an Empire) Picador 2007
22 David Isby (Afghanistan-Graveyard of Empires) Pages 64-65 Pagasus Books 2010
23 Magnus/Nady (Afghanistan)
24 ibid Page 117
25 James Michener (Caravans) Random House 1963
26 ibid Page 230
27 Isby (Afghanistan)
28 Magnus/Nady (Afghanistan) Page 122
29 ibid Page 36
30 Adamec (Historical Dictionary of Afghanistan) Page 328

31 Magnus/Nady (Afghanistan) Page 62
32 ibid Page 122
33 ibid Page 146
34 Gregory Feifer (The Great Gamble The Soviet War in Afghanistan) Page 55 HarperCollins 2009
35 ibid Page64
36 ibid Pages 59-82
37 Magnus/Nady (Afghanistan) Page 117
38 George Crile (Charlie Wilson's War) Atlantic Monthley Press 2003
39 Magnus/Nady Page 133
40 Steve Coll (Ghost Wars) Pages 289-300 The Penguin Press 2004
41 ibid Pages 332-333
42 ibid Page 79
43 ibid Pages 154-155
44 ibid Pages 154-158
45 ibid Pages 574-575
46 Michael Griffin (Reaping the Whirlwind-Afghanistan, Al Qa'ida and the Holy War) Pages 243-262 Pluto Press 2003
47 ibid 227
48 Gary C. Schroen (First In-An Insider's Account of How the CIA Spearheaded the War on Terror in Afghanistan)
49 Dalton Fury (Kill Bin Laden) St Martin's Press 2008
50 Pete Blaber (The Mission, the Men, and Me) Pages Berkley Hardcover 2008
51 LTC Troy D. Lewis (Training the Afghan Army-A Monograph) School of Advanced Military Studies-USCGSC, Fort Leavenworth, KS 2011
cgsc.contentdm.oclc.org/utils/getfile/collection/p4013coll3/id/2706
52 James F. Dunnigan and Albert A. Nofi (Dirty Little Secrets of the Vietnam War) Page 218 St Martin's Press 1999
53 U.S. Army Center for Military History 2010 (War in the Persian Gulf) Pages 24-26
www.history.army.mil/html/books/070/70-117-1/index.html
54 CSM Brian O'Conners (Correspondence with Author) 2013
55 author's files-draft TDA
56 author's files-email
57 Bridge info
58 author's files (PDSS PPT Briefing)
59 author's files (124 RTI MFR Report)
60 authors files (Risk Management Worksheet)
61 James Christ (The Bone Yard) Battlefield Publishing-Amazon DS 2011
62 author's files-(memo to CO)
63 author's files (wish list letter)
64 author's files (Mobe Manning Doc)
65 author's files (TF Martin history)
66 author's files (TDA PPT)
67 author's files-(ANA-TAG COMMAND PHILOSOPHY).
68 Rudyard Kipling (Kim) Page 118 Dover Thrift Edition 2005
69 Web: http://wikimapia.org/3538812/Camp-Phoenix

70 Troy D. Lewis (Training the Afghan National Army) Page 23 S.A.M.S. USAC-GSC, Fort Leavenworth, KS
71 Donald P. Wright Ph.D.et al (A Different Kind of War) Page 232 Combat Studies Institute Press-U.S. Army Combined Arms Center, Fort Leavenworth, KS 2010
72 ibid Page 231
73 John Masters (A Bugle and a Tiger) Viking Press, New York 1956
74 COL. Randy Davis (Correspondence with author)
75 ibid
76 ibid
77 Web: http://www.isaf.nato.int/history.html
78 CSM Brian O'Conners (Correspondence with author)
79 CSM Brian O'Conners (Correspondence with author) 2013
80 ibid 2013
81 ibid 2013
82 ibid 2013
83 ibid 2013
84 author's files (Memos to CO 2-11 Jan 04)
85 author's files (Memo to Major M 6 Jan 04)
86 Presidio Press 2003
87 author's files (memo 6 fEB 04)
88 author's files (memo 15 Feb 04)
89 author's files (memo to BG Mancino 4Feb 04)
90 Military Bookshop 2010
91 COL Randy Davis (Ret) (Correspondence with author)
92 ibid
93 ibid
94 author's files (MFR dated 28 Mar 04)
95 author's files (video clip)
96 author's files (Memo 3 Apr 04)
97 James Michner (Caravans) New York: Ballantine Books 1963
98 author's files (document)
99 author's files (memo to CO 11 May 04)
100 author's files (document 24 May 04)
101 author's file (medical records)
102 Michael Griffin (Reaping the Whirlwind-Afghanistan, Al' ida, and the Holy War) Page 23 Pluto Press 2003
103 Lester W. Grau (THe Takedown of Kabul-An Effective Coup de Main) Leavenworth Combat Studies Institute 2002
104 Michael Friscolanti (Friendly Fire-Tje Untold Story of the U.S. Bombing that Killed Four Canadian Soldiers in Afghanistan) John Wiley and Sons, Canada 2005
105 Ludwig W.Adamec (Historical Dictionary of Afghanistan) Page 54 Scarecrow Press 2012
106 author's files (memo to CHOPS 15 Jun 04)
107 Donald P. Wright Ph.D et al (A Different Kind of War) Page 254 Combat Studies Institute Press-U.S. Army Combined Arms Center, Fort Leavenworth, KS 2010
108 COL Randy Davis (Ret) (Correspondence with the author) 2013
109 CSM Brian O'Conners (Corresponence with the author) 2013

110 COL Randy Davis (Ret) (Correspondence with author) 2013
111 Donald P. Wright Ph.D et al (A Different Kind of War) Page 83 Combat Studies Institute-U.S. Army Combined Arms Center, Fort Leavenworth, KS 2010
112 Diane Publishing 1996
113 U.S. Army (FM 101-5 Chapter 5) 1997
114 U.S. Army (FM 100-14) 1998
115 Pashto for the bread [and or a mea] of Afghanistan and elsewhere-also known as Nan
116 J. R. R. Tolkin 1937
117 U.S. Army (AR 600-8-22) Page 50-52 1995 - 2006
118 U.S. Army (AR 670-1) Page 239
119 The ARCOM is also known as 'the Green Weenie' in the barracks
120 Major Max Krupp (Correspondence with author) 2013
121 This sort of controversy may have contributed to a DA message ALARACT 055/2007 dated 3 FEB 05 which allowed units smaller than companies to wear the patch of the unit they fought with.
122 John Krakauer (Where Men Win Glory) Page 315 Anchor, Rev Rep 2010
123 2nd Timothy 2:5
124 Military Job Description-Usually requires completion of a service school that awards a particular MOS
125 Use or lose it in next year's budget
126 The Holy Bible, New Testament -Acts 17:26
127 Certificate of Release or Discharge from Active Duty
128 Statistics provided by LTC (Ret) Randy Davis
129 Statistics provided by Texas Military Department
130 Basic Training Schedule provided by LTC (Ret) Randy Davis

Index

Made in the USA
Monee, IL
04 November 2026

67378304R00164